Disguised in Deception

E. A. Hargrave

DEDICATION

To my adorable darling Allie—to you, I owe my inspiration, my reason for a more promising future.

Table of Contents

The Value of Disguised in Deception

What you're about to read is information I have gathered from many public sites that will help the reader make better decisions regarding their health and that of their family. Although some of this information might not be applicable to your location, it will act as a guide for you to do some of your own research. You might not agree with some of the information I have found and shared in this book, but I would strongly encourage you to look for the information yourself.

Introduction

Disguised in Deception will give you a list of the most common ingredients in many of the grocery items sold in the USA. This book informs you about the ingredients and what they are used for, and some to avoid altogether. They are listed by their scientific and common names.

I have spent two years gathering information from many sources to bring you the most up-to-date information about what is really in your food. Some of these ingredients may cause long-term damage to you and your family. I will also show you how to get the toxic ingredients out of your body to become healthy once again.

2rd Edition

Section 1
Saving on Grocery Shopping

To start, here are some fast tips for how to save money. Some are merely common knowledge—however, I find that many people just plainly don't do them.

Compare prices and save!

People don't have the time to compare prices at other stores, however there are more effective ways you can manage this. One way is to look online. There are websites that will help you find the best deals and coupons in many states around the USA and the world. A couple of good websites to start off with are mygrocerydeals.com, the Retailmenot app, and mambosprouts.com. Stores now price match their competitor's products, all you have to do is find a store in your area! Some that I know of are Walmart, Target, Sears, Lowe's, and Kohl's. I would call ahead or look at their website for more information on their rules regarding price matching. I know there are several hundred other stores in the country that will price match. All you have to do is look!

Next, it's time to plan the menu.

This one is easy, but many people don't really like to do it. By planning out your meals, you only purchase what you need at the store. By having planned out your food menu over a week or two, it is much easier to shop.

Eat something before you go grocery shopping.

Have you ever walked outside and smelled a barbecue or McDonald's? All of a sudden, you're craving whatever you smell! When you shop while you're hungry, you'll find that everything in the store looks good. You end up getting food that you wouldn't normally buy, and this can end up costing you much more than if you had eaten a good, filling snack prior to your shopping trip.

Use coupons when grocery shopping.

Use coupons! They can be found almost anywhere. Many companies will accept coupons and better yet they might even have a coupon app like Wholefoods app, so you can scan them right from your smartphone! It is no longer necessary to troll over newspaper after newspaper to find exactly what you need. Here is another benefit—use them to price match! Contrary to popular belief, there are many stores that also use coupons for organic food, so if you're big into organic, this is the thing for you!

Be aware that not all brands are created equal.

Many people believe that name brands are always the best, but that'sn't actually true in most cases. The store brand is usually the same quality or sometimes even better. There are many websites that have more information on store brand vs. name brand.

Understand the unit price.

Calculating per pound price is not for everyone, but by comparing the unit prices, it will give you a better understanding of the best deal.

It's an impulse buy.

Everyone buys on impulse, and that's okay, but if you're seeking to reduce the cost of your shopping, this is something you need to adjust. Understand that just because the item is on sale doesn't mean you have to purchase it. Stick to the shopping list, and only buy what you went to the store for.

Organize your refrigerator.

Did you know that you can prolong the life of your food? All it requires is for you to properly arrange your food in your refrigerator. Keep fruits and vegetables in separate drawers, since fruits will most likely go bad faster than the veggies. Many people are not aware that some vegetables require humidity to remain fresh. Next time you're at your local market, notice that they spray water on the raw vegetables to maintain freshness.

Section 2
Shopping Local

In this section, we won't be talking about the reason why people should shop locally or why we should be keeping the money in our local economy or even how we can reduce the cost on the environment. No! In this section, we'll talk about the effect food cost has on your wallet, which ultimately affects your wellness. When I talk about shopping locally, I'm talking about farmers markets and local ma-and-pa shops. Most locally owned stores and farmers markets will provide better-quality food and even many other non-food-related items. Let's have a look at some reasons why buying at the local farmers market is better than buying at the large food chains.

Organic fruits and veggies.

GMO-free and organic fresh fruits and veggies can be found at local farmers markets, and are often cheaper than in the supermarket. You can always ask if you're not sure if a product is organic or GMO-free while you're shopping, it's easier to get better information face-to-face. Not all farmers grow GMO-free and 100% organic food, so by asking specific questions, you will have better decision-making capabilities.

It's great to eat seasonally.

When you shop at farmers markets, you're getting the freshest food, and because you're eating seasonally, you can taste the difference! Not only is the food fresh, ripe and tasty, did you know major supermarkets pick fruits and veggie far too early, which decreases the overall liveliness.

Safer foods.

Nutrients in foods from the large industrial processing plants can be polluted in many ways. In recent news, a food-processing company out of California had to recall nine million pounds of meat! This number was so high because they mass-produced the food. The preservatives in food are to help extend the lifespan of the product, but the same preservatives that are found in food can cause health risks to the people who consume them. Buying food from your local farmers market is usually safer and guaranteed for freshness. Unfortunately, this would be impossible in a large supermarket.

Better taste.

Food bought at local farmers markets just plainly tastes better! Of course, there are reasons for that too—by the time the food has made it to your table, it has had a chance to grow and mature all the way and it gathers all the nutrients that it needs to make your body healthy and strong. Supermarkets pick the food before it has a chance to mature, this not limits the nutrients, precious vitamins, and minerals. Fruits and veggies decay before they can even get ripe. You are eating food that has little to no nutritional value.

Section 3
Organic vs. GMO

What are GM foods?

Process of changing the genetics or altering the genetic makeup of any food crop or animal product is called "genetically modified." Foreign proteins are infused into food crops, making them genetically modified, the DNA of different plants are combined to create new species of foods. Other names used in describing these products include "genetically engineered" (GE) and "genetically modified organisms" (GMO). Soybeans, corn, canola (rapeseed) and cotton are the most common GMO crops. Most of these crops are either "insect resistant" or "herbicide tolerant."

There's a great deal of debate between organic and GMO foods, it's common to see GMO products also being called GE—this stands for "genetically engineered." These words and meanings are interchangeable, and mean virtually the same thing. The creation of GMO animal's the subject of controversy, Frankenstein's monsters are now being created! Consider genetically modified goats, this happened at a farm located at Utah State University—they mixed the spider gene with the goats' DNA, which produced milk with a silk protein, which is then removed and spun into silk thread.

The Daily Mail quotes Dutch bio-artist Jalila Essaidi as saying:

"Now, let's take this one step further. Why bother with a vest: imagine replacing keratin, the protein responsible for the toughness of the human skin, with this spider silk protein. This is possible by adding the silk-producing genes of a spider into the genome of a human: creating a bulletproof human. Science-fiction? Maybe, but we can get a feeling of what this transhumanistic idea would be like by letting a bulletproof matrix of spider silk merge with an in vitro human skin."

11

Even though this will stop a bullet fired at reduced speed, it was unable to stop a bullet from a .22 caliber gun shot at normal speed. .22 bullet at a normal rate is one of the requirements for today's bulletproof vests. The skin was on display at the National Natural History Museum Naturalis in Leiden, Netherlands, until Jan. 8, 2012.

Many genetically modified foods have a bacteria called (Bt), also known as "Bacillus thuringiensis." This bacterium is found naturally in dirt; it's lethal to insects. The Center for Food Safety estimates that 70% of processed foods contain genetically-altered elements. The first two (Bt) modified foods that were approved in the United States are potatoes and cotton. Research have shown that insects are building resistance to genetically modified crops and bio-pesticides that are now found in the human bloodstream.

Bacillus thuringiensis (Bt), when ingested, produces a crystalline protein that kills cells and makes holes in the insect's stomach, therefore killing whatever ingested it. The Monsanto Corporation is putting up to six different genetic material (Bt) genes and glyphosate-tolerant genes into one of their corn crops. She states **"Rather than prevent insect resistance, this approach will probably accelerate it,"** according to Shannah Schmitt, in Aug 29, 2011. In 2011, the EPA granted approval for Syngenta's Agrisure 3122, which is corn seed stacked with multiple GMO traits. According to the USDA, corn seed prices rose 146% from 1999 to 2011. GMO corn is about 86% of the nation's crops.

Science knows very little about the effects of consuming food with (Bt) genes. Short-term studies have exposed signs of toxicity in the kidney and liver. A 2009 study in the International Journal of Biological Sciences found a **"clear negative impact on the function of these organs in rats consuming GMO corn varieties for just ninety days."** - Shannah Schmitt on Aug 29, 2011.

Ask yourself this question. The study shows the negative impact after only ninety days, what could happen after a lifetime or generations of eating GMO and (Bt) corn? Research has shown that (Bt) toxins are showing up in the bloodstream of humans, researchers in Canada have found (Bt) in the blood of women. A study published in Reproductive Toxicology in February of 2011 found the (Bt) toxins in 93% of pregnant women's blood samples, 80% of fetal blood samples, and 69% of non-pregnant women's blood samples. This study shows that (Bt) from the mothers can be passed on to the unborn infant.

"Pigs fed a combination of genetically modified soy and corn suffer more frequent severe stomach inflammation and enlargement of the uterus than those who eat a non-GM diet, according to a new peer-reviewed long-term feeding study published Tuesday in the Organic Systems Journal. In pigs eating genetically modified crops, the average rate of severe stomach inflammation was nearly three times as high as that for other pigs (32 percent vs. 12 percent). Among male pigs eating a GM diet, the rate of severe stomach inflammation was four times higher." - Monica Eng, Tribune reporter, June 11, 2013

"The results indicate that it would be prudent for GM crops that are destined for human food and animal feed... to undergo long-term animal feeding studies, preferably before commercial planting, particularly for toxicological and reproductive effects," concluded Carman and her colleagues, who include Iowa-based farmer as well as crop and livestock advisor Howard Vlieger.

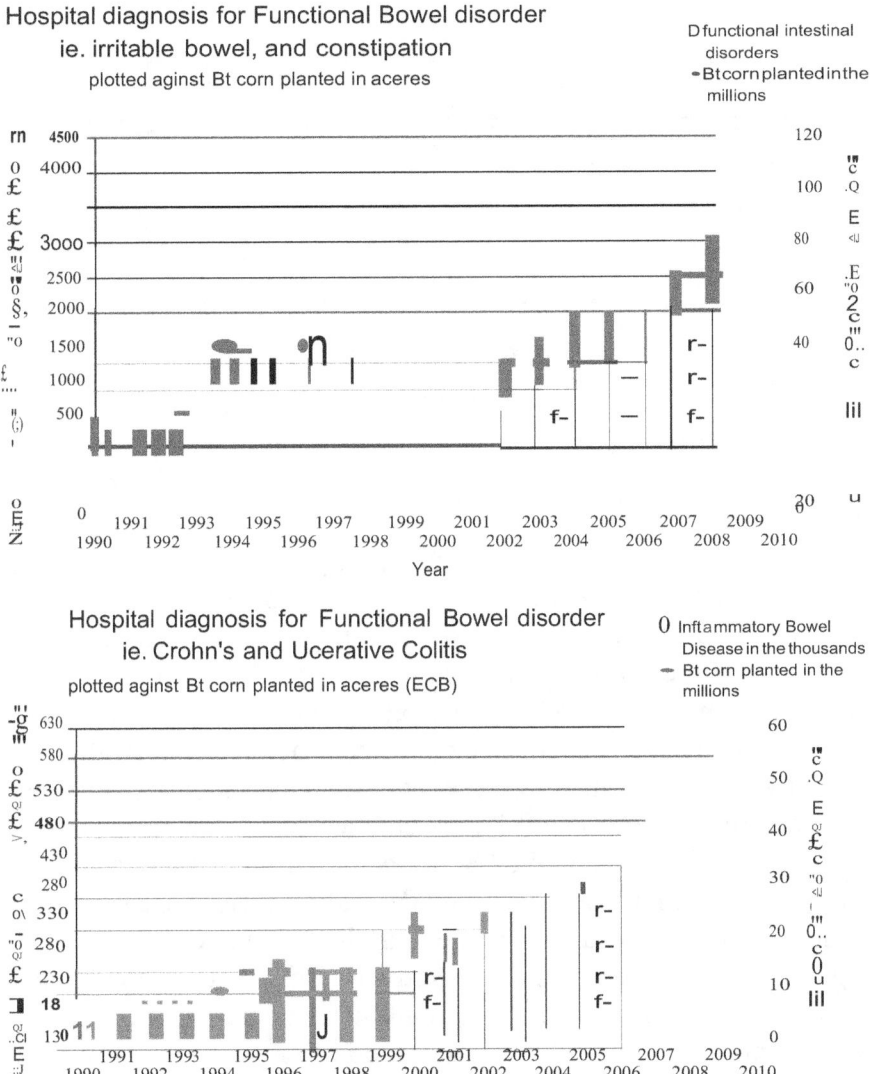

Hospital diagnosis for Functional Bowel disorder ie. irritable bowel, and constipation
plotted aginst Bt corn planted in aceres

D functional intestinal disorders
- Bt corn planted in the millions

Hospital diagnosis for Functional Bowel disorder ie. Crohn's and Ucerative Colitis
plotted aginst Bt corn planted in aceres (ECB)

0 Inflammatory Bowel Disease in the thousands
- Bt corn planted in the millions

14

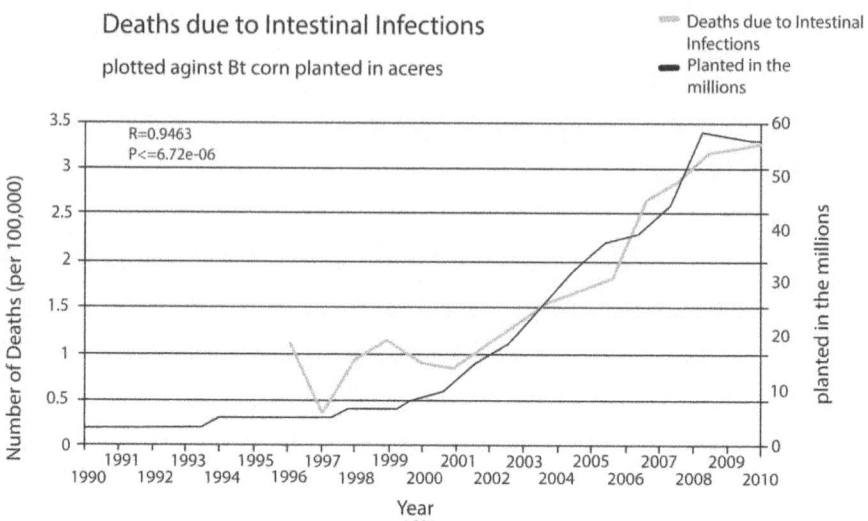

Deaths due to Intestinal Infections

plotted aginst Bt corn planted in aceres

Deaths due to Intestinal Infections

Planted in the millions

Herbicide resistance and pesticide production go hand in hand, as much as 86% of corn, 90% of all soybeans, and 93% of cotton are GMO varieties. Genetic modification is now used in more than half of all planted crops in the U.S. You're consuming genetically modified foods almost daily, unless you produce and grow all your food. Genetically modified crops—**"none of them are labeled"**—now include sweet corn, peppers, squash, zucchini, rice, sugarcane, **"canola oil,"** flax, chicory, peas, and papaya. Over 70% of the milk in the United States comes from cattle injected with a GM hormone or are eating GMO alfalfa. Bees use GMO pollen from crops, and even some vitamins include GMO ingredients. Some sources conservatively estimate that more than 70% of processed foods in the United States contain GMO ingredients because most processed foods contain corn or soy.

GMO foods are not labeled within the United States, because the biotech industry has convinced the FDA that genetically altered crops are no different from organically grown crops. The FDA, however, doesn't perform any independent testing for human or animal safety, and relies strictly on research produced by the makers of the genetically engineered crops. The main GMO producer in the USA, Monsanto, is the same manufacturer that made Agent Orange, and Roundup, the weed killer. Of course they would have a biased opinion of their own product!

Many nations around the world have laws in place that demand GMO products to be labeled, while some countries and regions have banned genetically modified crops and animals entirely. Some countries and regions have only banned certain chemicals and bacteria found within GMO-produced foods. A research study shown that GMO milk could be the cause for puberty starting earlier in girls, some young girls are experiencing puberty around the ages of eight and nine years old. Even in the last three decades, children (especially young women) have matured at younger ages—precocious puberty is ten times more common in young girls than boys. Just two generations ago, the average age of puberty was about fourteen to sixteen years old. Nowadays, girls start puberty around the age of seven.

Research point to the the added growth hormones found in genetically engineered crops and farm-raised cows. In my personal opinion, adding growth hormones to animals that we eat, cook with, and drink from will only pass on the growth hormones through our bodies. Early puberty can set the stage for emotional and behavioral problems, early puberty is tied to low self-esteem, depression, eating disorders, alcohol use, earlier loss of virginity, more sexual partners, and of course, an increase in potential of sexually transmitted diseases. The evidence suggests that early puberty can also cause diabetes, infectious diseases, and cardiovascular diseases, as well as cancer in young women.

Although early puberty is more prominent with young females, early puberty does affect young males as well. About thirty years ago, males' puberty used start between fifteen to seventeen years old, now, studies have shown that boys are starting puberty at the age of ten years old. Research has linked early puberty in young males with diabetes, infectious diseases, prostate cancer, and testicular cancer.

New York Times article reported, **"Animal studies show that the exposure to some environmental chemicals can cause** bodies to mature early. Of particular concern are endocrine-disruptors, like 'Xeno-estrogens' or estrogen mimics. These compounds behave like steroid hormones and can alter puberty timing. For obvious ethical reasons, scientists cannot perform controlled studies proving the direct impact of these chemicals on children, so researchers instead look for so-called 'natural experiments,' one of which occurred in 1973 in Michigan, when cattle were accidentally fed grain contaminated with an estrogen-mimicking chemical, the flame retardant PBB. The daughters born to the pregnant women who ate the PBB-laced meat and drank the PBB-laced milk started menstruating significantly earlier than their peers."

Research indicates that GMO foods may be responsible for obesity in children and adults. Various animal studies indicate some serious health risks linking GMO foods to infertility, immune problems, accelerated aging, insulin resistance, and changes in major organs and the gastrointestinal system. More and more health professionals and even allergists are recommending that their patients get on a GMO-free diet. Animal testing on rats has demonstrated that genetically modified food can cause tumors and possibly early death. States such as California have tried unsuccessfully to create laws that force GMO foods to be labeled. GMO giants rebut by stating, **"If they had to label the GMO products, that would increase the overall cost of the labeling, and increasing the monetary value for the consumer."**

Organic food is, by definition, supposed to be free of genetically modified ingredients. Organic crops are required to be isolated from other non-organic crops. GMO companies have blurred the lines of what is and is not organic. The new standard of organic foods is unchanged, unaltered, pesticide, herbicide, and growth-hormone-free. It is becoming increasingly harder to find 100% organic food. GMO plants cover approximately 80% of the United States. Bees and other insects, animals, and birds take genetically modified pollen from genetically altered plants to organic crops, which change over time from organic to genetically modified organisms. GMO plants and animals are not natural.

17

They are not even considered to be part of the known list of species or part of the food chain associated with the human diet.

Current USDA regulations allow food products that contain 95-100% certified organic components and require 100% organic feed for organic livestock to use the USDA organic seal. Products labeled "made with organic ingredients" require 70% of the ingredients to be organic, but 100% must be non-GMO. The new policy, passed on September 16th, 2013, stated that an exempt ingredient such as synthetic fertilizers, pesticide, sewage sludge, and GMO can be indefinitely permitted unless 2/3 majority of the National Organic Standards Board votes to remove the exempt man-made ingredients.

The law clearly states that these exemptions are authorized for a five-year period. This is to promote the development of innate (or organic) alternatives. The exemptions are needed by law to expire, this law is called "sunset." policies can only be reestablished by a two-thirds "decisive" majority vote on the National Organic Standards Board (NOSB) and include a public critique. While this is the law, the USDA has stated that it will no longer operate in this manner. The USDA organic seal is what a true organic product is allowed to use. The label signifies that the product is at least 85% organic. More information on stores that are free of GMO can be found at nongmoproject.org. More information on organic foods can be found at organic.org.

Natural vs. Organic

How the food is grown	Organic	Natural/Conentioal
No toxic persitient pesticides	Yes	No
No synthentic growth hormones	Yes	No
No petroleum-based fertilizers	Yes	No
No cloning	Yes	No
No GMOs	Yes	No
No sludge & irradiation	Yes	No
How the food is processed		
No toxic persitient pesticides	Yes	No
No artificial colors or flavors	Yes	No
Animal welfare requirements	Yes	No
Low levels of environmental pollution	Yes	No
Leagal restrictions on allowable materials	Yes	No
Certification required, including inspections	Yes	No

Many people don't know the difference between organic and natural foods. The food and GMO industries want you to believe that there's no difference, and that's simply not true. Natural doesn't mean organic or GMO-free. "Natural foods" are often misleading. Many believe they have little processing and don't contain any hormones, antibiotics, or artificial flavors. The FDA and the USDA have no requirements for companies to use the label "natural." Food manufacturers will often mislead consumers by placing a "natural" label on the package, even though many of these foods contain severely processed ingredients. Organic products are heavily regulated within the United States, certification inspectors are from a third party to ensure that the manufacturers are growing and processing organic products, or else they can get shut down and/or face a heavy fine.

Quote from organicitsworthit.org: **"It's easy to feel overwhelmed as you walk the aisles of the grocery store in search of products that are good for you, your family, and the planet. To simplify your shopping experience and insure that you're purchasing products that equip with your values, look** for merchandise with the USDA organic label. All products bearing this label must be made according to a strict set of government standards and regulations. Food manufacturers that use the organic seal are certified according to the United States Department of Agriculture (USDA) under the Organic Foods Production Act of 1990 (OFPA). "Natural," on the other hand, generally refers to food items that are not altered chemically or synthesized in any configuration.

Certification Agencies

In 1990, the US Congress passed the Organic Food Production Act (OFPA). The purpose of the OFPA was to ensure that all agricultural products were marked as organic so they would, in fact, meet consistent and uniform measures. The United States Department of Agriculture's National Organic Program (NOP) strives to continue to deliver on this objective. Every product that's sold in the United States under the pretenses that it is organic, whether it is foreign or domestic, is forced to adhere to the October 21, 2002, USDA's National Organic Program.

All products that are sold on United States store shelves that bare the organic label, "100% organic," or "made with organic elements" must be certified! Having the organic certification provides declaration that the product fulfills the ideals authorized by National Organic Program (NOP). These measures are comprehensive and encompass every aspect of organic farming, processing, transportation, labeling, and promotional material. The International Association of Natural Products Producers (IANPP) is attempting to get the "natural food" designation as clear as possible for people to understand. Nevertheless, they are not a certification body.

USDA labels for organic food have legal implications—a food company must encompass the detailed rules and regulations before they are allowed to give the USDA stamp. Natural labels are employed without restrictions by manufacturers due to deficiency of suitable guidelines. The other way to avoid GMO foods is to look for the **NON-GMO PROJECT VERIFIED** label.

Certified to ensure <u>no</u> GMOs, chemical fertilizers, synthetic substances, irradiation, sewage or sludge are used in the production of USDA organic items.	Verified by GMO testing, which assures that a product has been produced according to consensus-based best practices for GMO avoidance.

Getting a full list of genetically modified foods is nearly impossible to offer in the United States because of the fact, there aren't any laws for genetically modified crops. Some have estimated as many as 30,000 different products on grocery store shelves are "modified." Many processed foods contain soy, which is one of the most common GMO crop in the USA.

The list of the most popular GMO's in the U.S. are as follows:

Dairy Products:

Close to 22% percent of cows in the U.S. are injected with recombinant (genetically modified) Bovine growth hormone (rBGH).

Cottonseed Oil:

Did you know that cottonseed oil is in many of your favorite foods? It might be a shock to some knowing that cottonseed oil is used in salad oils, mayonnaise, salad dressings, many sauces are made with cottonseed oil. The use of cotton as a food product is nothing new, in fact, cotton was a major ingredient of the American diet until the 1940's. Cottonseed oil is still used as a cooking oil in frying for commercial and home cooking. It is an additive to shortening and margarine because of the lack of color and very light taste, it has also been a favorite in baked goods and icings on cakes.

Corn:

There are 19 GM corn varieties that we know about. Studies on the just 3 of them have shown when rats who ate the GM corn developed hormonal imbalances detected along with other compounds in the blood and urine. The findings suggest that each of the 3 corn varieties tested MON 810 and MON 863, which are resistant to pests, and NK 603, which is modified to withstand weed killer. The conclusion of the study, female rats fed MON 863, had high blood-sugar levels and very high concentrations of triglycerides. These symptoms indicate the early stages of diabetes.

From the second study, they tested the effects of GM corn over a 30, 60, and 90 days duration, in each case it had shown to modify the enzymes and damage do their genes. Shockenly, the rats had cell damage and the rate of DNA fragmentation had increased! Some of the GM proteins have been found to transfer to humans and animals which caused health problems.

Flax:

GM flax was all but forgotten, a FP967 strain of flax has assumed to be found in 30 countries. This illegal form of GM flax was made in the late 1980s, by the Crop Development Center in Saskatoon, Saskatchewan, the name was changed to Triffid later on. In the 1990s the crop was approved for commercial usage in Canada, later on the U.S. Consumers were concerned about the safety of this modified flax eventually used in the commercial market died until now! FP967 has been found in flax stock to this day, showing that it is virtually impossible to stop GMOs. GMO's once released into the environment cause lasting health and environmental concerns.

Honey:

Bees by instinct seek out flowers and make honey, the bees do not know what flowers are genetically modified. Bees have been found to contain extremely high amounts of toxic insecticides about "700,000 times a bee's lethal dosage," have been found in genetically modified crops. The biggest concern is the toxin clothianidin, this chemical is made by Bayer Corp. Clothianidin is absorbed by the plant's vascular system, the pollen and nectar release by the plant are containing a cocktail of harmful chemicals.

The use of clothianidin on GM crops and seeds date back to 1991 and became widely used in 2003. The Honey made from the nectar is extremely poisonous, the effects of insecticides containing clothianidin have an effect on the human brain causing neurological damage.

Meat:

Meat and dairy products usually come from animals that have eaten GM feed.

Peas:

GM peas have a gene that came from a kidney bean. The genetic transfer of this gene into peas is associated with poor immune responses in animal testing. This reaction in mice would suggest that this gene could also produce serious allergic responses in people.

Potatoes:

Studies conducted on GM potatoes have shown that potatoes produce their own insecticide, the rats who consumed the GM potatoes had exhibited signs of cell growth in their digestive tract showing early stages of cancer. The consumption of GM potatoes **"Inhibited development of their brains, livers and testicles, partial atrophy of the liver, enlarged pancreases and intestines and immune system damage."** Gurian-Sherman said the USDA had failed to guarantee a "rigorous" analysis of the crop's possible consequences, and added: **"We simply don't**

know enough about RNA interference technology to determine whether GE crops developed with it are safe for people and the environment." said Doug Gurian-Sherman, Ph.D., CFS director of sustainable agriculture and senior scientist.

Papaya:

The first virus resistant papayas were commercially grown in Hawaii in 1999, Transgenic papayas now cover nearly three-quarters, of the total Hawaiian papaya crop. Monsanto donated technology to Tamil Nadu Agricultural University, Coimbatore, for creating a papaya resistant to the ringspot virus in India. There are ways to identify a GM papaya over a non-GMO papaya, some of the identifying traits of a GMO papaya are on the labeling. If the label states "Rainbow," "Sunrise," or "Strawberry" papayas, the fruit bearing these names are of the hybrid variety bred with the GMO variety. A non-GMO papaya has a rich and deep yellow skin, the skin should never be golden or pink, these colors are the tall-tail signs of the GM hybrids, try and look for Organic "Kapoho Solo" or Kapoho" these varieties are unchanged non-GM hybrids.

Another way to identify GMO papaya is to look at the number on the sticker/label, the PLU code is the best way to tell if the fruit is GM or Organic.

 • Conventionally grown: Four numbers, ie. "1034"

 • Organically grown: Five numbers, starting with the number 9 "90123"

 • Genetically Modified: Five numbers, starting with the number 8 "80123"

Rapeseed:

Also known as Brassica napus, is commonly known as canola oil. The danger of rapeseed is that over 90% of our supply is genetically modified. The media proclaims that rapeseed is a heart healthy oil. Most people don't know the history of repressed, began during the Industrial Revolution. Rapeseed oil was used as lubrication for ships and steam engines, which was ideal for wet metal. In 1995 a major breakthrough technique was created that altered the DNA in rapeseed allowing it to become resistant to toxic herbicide. The monounsaturated fatty acids in rapeseed oil are erucic acid, this form of fat and acids are associated with Keshan's disease, leading to the possible formation of fibrotic lesions on the heart.

Rice:

Did you know that rice contains human genes? The biotechnology company Ventria Bioscience, is currently growing this rice in Junction City, Kansas. The GM rice is spliced with genes from human livers, for use in pharmaceuticals. This approach is turning the number 1 grown grain in the world into a drug for profit. The 3,200 acres of human gene modified rice is not as of yet, approved for human use. The open planting of GM crops has in, fact, infected other plant species. Another issue with rice is the high amounts of arsenic found in the rice. The levels are even higher in infant foods.

Red-hearted Chicory (Radicchio): Scientists have created a strain of Chicory that has a gene that makes males sterile.

Soybean:

Currently, genetically modified soybean are grown on 93% of soybean farms. Soy foods include, soy beverages, tofu, soy oil, soy flour, and lecithin. Research indicates that only 3 generations of hamsters, fed a GM soy based diet lost the ability to reproduce. Shockingly 25% of the 3 generation had died younger than previous generations! Even if you never went to the store and picked up some soybeans or soy milk, you are still eating soy somehow.

Cattle are fed soy, which in turn ends up in the meat we all consume. Soy is used in cooking oil, baby formula, and in thousands of other food products. Studies have found glyphosate and aminomethylphosphonic acid (is a metabolite of glyphosate), or AMPA. The levels found in the GM soy even by their own makers Monsanto characterized as "extreme."

Sugarcane:

Sugarcane was made resistant to some certain viruses, insect, bacterial, and also tolerance to herbicides. Sugar cane was modified to increase the sucrose yields. High levels of sucrose in the U.S. diet could be associated with PCOS and diabetes. When the body consumes sucrose it turned into glucose, as a result the insulin in the body increases. Years and years of high insulin result in damaged insulin receptors, causing chronically high blood glucose, this could lead to type 2 diabetes.

Sugar Beets:

Monsanto released GM sugar beets in march 2005, About 50-60% of all sugar in the US is made from sugar beets, to this day, about 90% of sugar beets are the Monsanto's GM crop.

Squash:

Some zucchini and yellow crookneck squash are also GMO. Check the sticker to be sure.

Tobacco:

GM tobacco is designed to kill pests by destroying the digestive tract, in a very similar way that Bt corn does. Smokers of GM tobacco are slowly killing themselves and in America it's legal to "slow kill" the population, they can get away with this tactic by law as long as the consumer know the risk they are allowed to knowingly kill! The changes made to tobacco are nicotine content, molecular pharming, herbicide tolerance, resistance to nematodes and viruses, this is linked to male sterility. A new study conducted at the Colorado School of Mines in Golden, Colorado, have found 3 pesticides that were never even known to be in cigarettes.

The 3 pesticides found in tobacco are:

- **Flumetralin:** This carcinogen is known to be extremely toxic to humans, it is forbidden in the UK for use in or on tobacco. Flumetralin is a known endocrine disruptor.

- **Pendimethalin:** Known to target the thyroid gland and known to be an endocrine disruptor.

- **Trifluralin:** Yet another endocrine disruptor, this toxic substance attacks the glands in the body affecting the hormones and is linked to breast and prostate cancer.

Tomatoes:

Have been constructed for longer shelf life used to prevent bacteria and funguses that can cause tomatoes to rot and degrade. Research conducted on the safety of GM tomatoes, some of the animal subjects had died soon after they started eating the GM tomato.

Vegetable Oil:

The majority of vegetable oils and margarine that are sold in the United States are being mixed with soy, corn, canola (rapeseed), and cottonseed oil. All of which are GM and are tied to health conditions. Even if you were to buy these oils in the organic form, some are still may not be healthy for you.

Other Ingredients Derived from GMO:

Amino Acids, Aspartame, Ascorbic Acid, Sodium Ascorbate, Vitamin C, Citric Acid, Sodium Citrate, Ethanol, Flavorings ("natural" and "artificial"), High-Fructose Corn Syrup, Hydrolyzed Vegetable Protein, Lactic Acid, Maltodextrins, Molasses, Monosodium Glutamate, Sucrose, Textured Vegetable Protein (TVP), Xanthan Gum, Vitamins, Yeast Products.

Section 4
Local Butchers vs. Supermarket Butchers

Consider shopping at local butchers' over supermarkets, it's true that there are bad butchers out there, some are with large supermarket chains and some have their own shops. You might benefit from doing some digging and getting informed about your local butcher shop around town.

Many farmers take their livestock to local butchers mainly because farmers receive more money for the cattle. The meat is fresh and have less; fillers that increase the cost. Local butchers have to compete with large food chains, so they are more willing to make the cuts you ask for. Local butchers are also helpful for giving advice on how to cook and prepare the cuts of meat. Not all, but some, of the local butchers are USDA organic approved and certified. This could mean their meat have no growth hormones and is non-GMO, you should ask.

Most news outlets and online forums haven't talked about the one big advantage local butchers have over supermarket chains. That advantage is safety! In February of 2014, Rancho Feeding Corporation in California recalled nine million pounds of beef, this was due to, **"diseased and unsound animals,"** which is a Class I recall status.

Definition for Class I: **"This is a health hazard situation where there is a reasonable probability that the use of the product will cause serious, adverse health consequences, or death."**

When you shop at a local butcher's the livestock is not mass produced, which means that more health and safety measures are easily followed. The fact is, supermarket chains have GMO in their food which may not be well regulated, and that includes growth hormones. Another issue is how most companies kill their livestock. Most of the time, it's not in a humane way, and knowing that the food you're eating lived humanely and died humanely can be easier to accept. Nearly every large company that supplies livestock for supermarkets has, at one time or another, been in the news for poor treatment of their livestock.

The full list can be found here: mercyforanimals.org

Section 5
Introduction to Allergies

I want everyone to ask themselves these questions—where did allergies start, and why? Have you ever asked older generations if they know of anyone who had a food allergy or medical conditions like peanut allergies, PCOS, or even diabetes while growing up? In my personal opinion, the number-one problem we're having, is that more and more generations are developing health problems due to the food and the environment. In this section, we'll look at allergies and medical conditions that might be linked to food, and the history behind them.

An allergy is **"an abnormally high sensitivity to certain substances."** An allergy happens when the body is exposed to a new organism, the body will produce antibodies. If for some reason the body can't make the right antibodies, the result is an allergy.

So then why do some people have them, and why do some countries have virtually none at all? According to the Centers for Disease Control and Prevention allergy statistics: **"Food allergy is a growing public health concern. As many as 15 million people have food allergies. An estimated 9 million, or 4%, of adults have food allergies. Nearly 6 million, or 8%, of children have food allergies, with young children affected most. Boys appear to develop food allergies more than girls."**

Have you ever wondered why so many children "nowadays" have allergies to peanuts? In the early 1900s, allergic reactions to peanuts were very rare and not spoken about, mainly because the allergy was so uncommon. Today, 1.5 million children in the USA are allergic to peanuts. Peanut allergies have suddenly become the number-one cause of death from a foodborne allergy, being in a class of allergens that have the ability to trigger an anaphylaxis effect. These circumstances are accompanied by the risk of an asthma attack, shock, respiratory failure, and even death.

Peanut allergy soon reached epidemic proportions—1.5% of a million children.

These numbers fit the true definition of an epidemic, even though that word has never been used in mainstream news, for fear of the reaction the public. Research indicates that food allergies may be a trigger associated with other allergic conditions, such as atopic dermatitis and eosinophilic gastrointestinal diseases. Even though allergies that developed in childhood, such as milk, eggs, wheat, and soy, generally originate in childhood. They appear to become a lifelong allergy.

A survey published in 2008 by the Centers for Disease Control and Prevention, **"…about an 18% increase in food allergy was discovered between 1997 and 2007."** Centers for Disease Control and Prevention published in 2013, **"…food allergies among children increased approximately 50% between 1997 and 2011."**

In an online report from the website *NaturalNews*, they stated:

"As early as 2008, NaturalNews.com reported about a condition called Morgellon's disease. The article went on to report the symptoms of the disease as follows: crawling, stinging, biting and crawling sensations; threads or black speck-like materials on or beneath the skin; granules, lesions. Some patients report fatigue, short-term memory loss, mental confusion, joint pain, and changes in vision. Furthermore, there have been reports of substantial morbidity and social dysfunction leading to a dip in work productivity, job loss, total disability, divorce, loss of child custody, and home abandonment."

According to the report from *Natural News*, the finding of Morgellon's disease was regarded as a hoax. But additional investigations, found out that Morgellon's disease was, in fact, real! And may have links to genetically modified food.

This is a quote from NaturalNews.com: **"Despite this link being established, the CDC declared Morgellon's disease of** unknown origin. Worse, the medical community could not offer any information to the public regarding a cause for the symptoms. When the fiber was broken down, and its DNA extracted, it was discovered to belong to a fungus. Even more surprising was the finding that the fibers contained Agrobacterium, a genus gram-negative bacteria with the capacity of transforming plant, animal, and even human cells. Morgellon's disease is not the only condition associated with genetically modified foods. A growing body of evidence has shown that it may cause allergies, immune reactions, liver problems, sterility, and even death. Moreover, based only on the human feeding experiment conducted on genetically modified food, it was established that genetic material in a genetically modified food product can transfer into the DNA of intestinal bacteria and still continue to thrive."

Most food allergies have been linked to GMO foods. People should do research on their own medical conditions, you may find that your medical conditions are associated directly or indirectly with GM foods. Genetically altered foods should be one of the first things that a person who is suffering from allergies should look into. This should be a concern for those specific people because GM foods are not tested, regulated, or required to be labeled with the extra additives that may be triggering a reaction.

Why should we be concerned about GMO foods?

The evidence points towards GM foods cause allergy risks. Currently, the list of GM food products intersects with the eight most common food allergens: eggs, milk, fish, peanuts, shellfish, soy, tree nuts, and wheat.

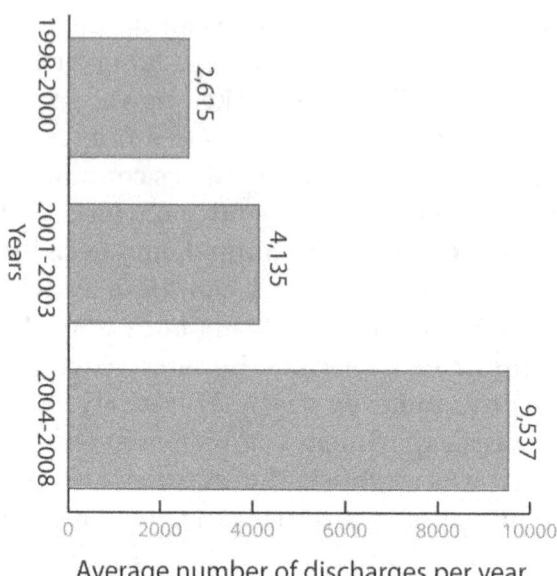

Average number of discharges per year

Over the last ten years, 1 out of every 17 children in the U.S. have acquired a food allergy; across the nation, there has been an increase of 265% in food allergy emergencies since 2004. Some proteins in food are not natural. In studies, GMOs have been linked to triggering most allergic reactions in people. Foreign proteins converted and added to foods, these proteins have never been eaten by humans before in history, or even tested for their long-term safety.

Section 6
Cancer

The best way to explain cancer is where the cells in the body, for some reason, have an allergic reaction and grow abnormally. There are many types of cancer, they all begin because of an uncontrolled growth of cells within the body. Cancer is the second-leading cause of death in the United States. Half of men and one-third of women in the United States will develop cancer during their lifetime.

The world's first documented case of cancer was in ancient Egypt. In 1500 B.C., a researcher found a papyrus in Egyptian texts with detailed descriptions of a tumor in the breast, which was determined to be the cause of death. Cancer has been around for centuries! Since it's nothing new, a question one should ask would be, why does it seem to be getting worse?

Why do more and more people seem to be falling ill to different types of cancer no one has heard of? Some might dispute that modern medicine is making it easier to diagnose cancer. As of Mar 22, 2017 - In 2016, an estimated 1,685,210 new cases of cancer will be diagnosed in the United States will have cancer, according to cancer.org. I don't discount that the number of humans living on earth now, versus the past, however, this would make the number of cancer survivors much more than previous years.

Five-Year Cancer Survival Rates
(All Types of Cancer)

— Men
Women

Nations

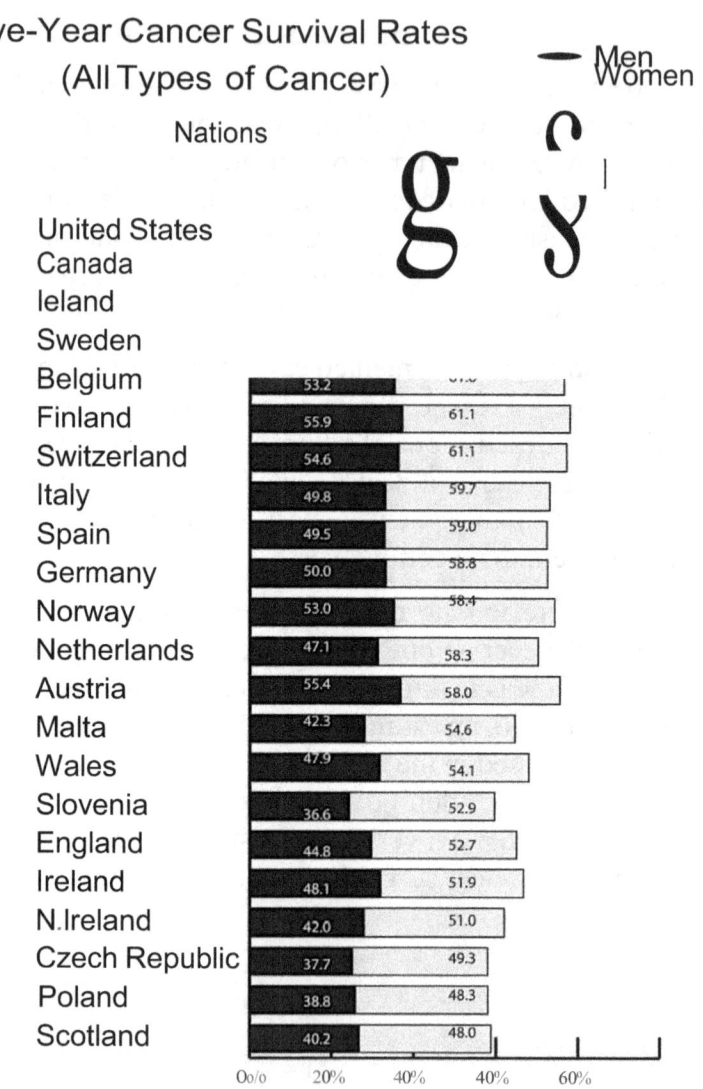

Nation	Men	Women
United States		
Canada		
Ieland		
Sweden		
Belgium	53.2	
Finland	55.9	61.1
Switzerland	54.6	61.1
Italy	49.8	59.7
Spain	49.5	59.0
Germany	50.0	58.8
Norway	53.0	58.4
Netherlands	47.1	58.3
Austria	55.4	58.0
Malta	42.3	54.6
Wales	47.9	54.1
Slovenia	36.6	52.9
England	44.8	52.7
Ireland	48.1	51.9
N.Ireland	42.0	51.0
Czech Republic	37.7	49.3
Poland	38.8	48.3
Scotland	40.2	48.0

0o/o   20%   40%   40%   60%

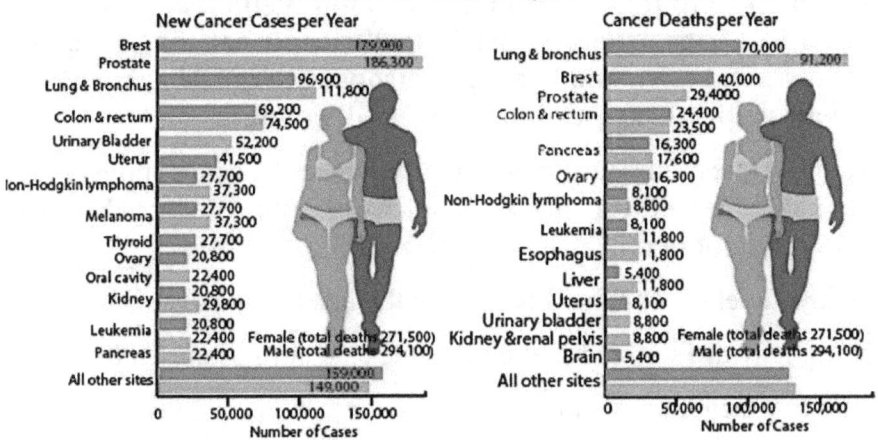

Cancer Incidence and Mortality in the United States

New Cancer Cases per Year

Brest	179,900
Prostate	186,300
Lung & Bronchus	96,900 / 111,800
Colon & rectum	69,200 / 74,500
Urinary Bladder	52,200
Uterur	41,500
Ion-Hodgkin lymphoma	27,700 / 37,300
Melanoma	27,700 / 37,300
Thyroid	27,700
Ovary	20,800
Oral cavity	22,400
Kidney	20,800 / 29,800
Leukemia	20,800 / 22,400
Pancreas	22,400
All other sites	159,000 / 149,000

Female (total deaths 271,500)
Male (total deaths 294,100)

0   50,000   100,000   150,000
Number of Cases

Cancer Deaths per Year

Lung & bronchus	70,000 / 91,200
Brest	40,000
Prostate	29,4000
Colon & rectum	24,400 / 23,500
Pancreas	16,300 / 17,600
Ovary	16,300
Non-Hodgkin lymphoma	8,100 / 8,800
Leukemia	8,100 / 11,800
Esophagus	11,800
Liver	5,400 / 11,800
Uterus	8,100
Urinary bladder	8,800
Kidney &renal pelvis	8,800
Brain	5,400
All other sites	

Female (total deaths 271,500)
Male (total deaths 294,100)

0   50,000   100,000   150,000
Number of Cases

If population is one of the biggest factors in cancer rates, then why do China, Japan, and India have lower cancer rates than the USA? Yet their population dwarfs that of the United States? From an article by Michelle Castillo on Feb 4, 2014, **"By 2030, the number of people worldwide who are diagnosed with cancer is expected to skyrocket to 21.6 million, a high 53 percent increase from the latest stats reported in 2012,"** the World Health Organization reported.

In an article from *The Guardian*, stated, **"Cancer cases worldwide are predicted to increase by 70% over the next two decades, from 14m in 2012 to 25m new cases a year."** World Health Organization. The latest World Cancer Report says, **"It is implausible to think we can treat our way out of the disease and that the focus must now be on preventing new cases. Even the richest countries will struggle to cope with the spiraling costs of treatment and care for patients, and the lower income countries, where numbers are expected to be highest, are ill-equipped for the burden to come."** Globally cancer rates have increased from 12.7m in 2008 to 14.1m in 2012, the death toll was estimated at 8.2m. It's expected to reach 25m a year over the next 20 years reaching a 70% increase.

The Guardian, cancer rates in North America and Western Europe are much higher than that of Africa. **"The incidence rate is rising fast in the developing world, but is still markedly lower in Africa, where 88 per 100,000 people got cancer, than in North America and Western Europe, where 334 and 335 people respectively per 100,000 were diagnosed."**

Another question you should ask is, why are the cancer rates getting higher if the medical technology is getting better? Cancer research is more funded than ever before!

Ask yourself—what has changed in the last few years?

Well, here are some of the reasons the cancer rates are on the rise.

- Genetically modified foods

- Fluoridated water

- Pesticides on produce

- Herbicides on produce

- Vaccines

- Tobacco

In 1971, Nixon introduced the war on cancer, spewing out mass research, which was funded and backed by the U.S. government. This escalated until the 1980s, when over $50 billion a year was spent to "find the cure." Thanks to the U.S. government's own statistical abstracts, we have finally discovered the truth. One of the most effective ways to clearly understand cancer data is by looking at the cancer death statistics versus the cancer survivor statistics. The truth can be found by looking at the cancer death rates. Many people can't afford to fight cancer, or even have an idea that they have it. In epidemiology (a science that investigates the presence or absence of diseases and disorders in defined populations), deaths from the disease are always measured in deaths per 100,000 populations.

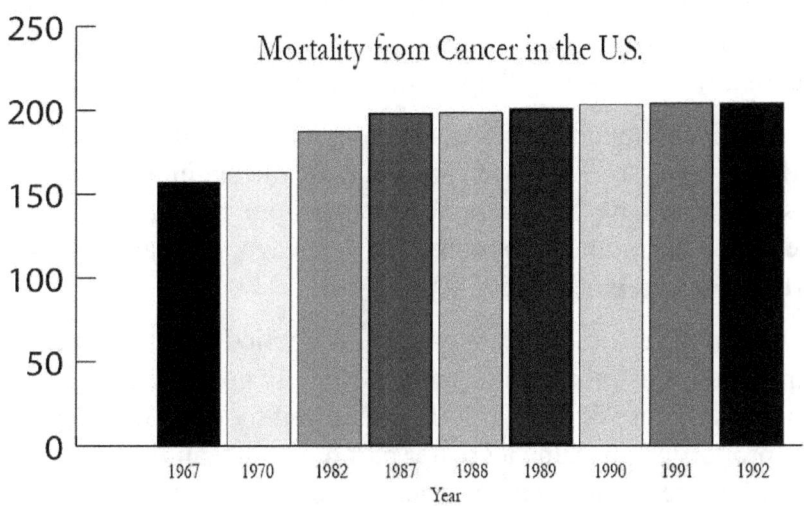

Independent examination by the CA Journal for Cancer Clinicians, Jan 97, put the 1993 death rate at 220 per 100,000. In the year 2000, the overall rate had climbed to 321 per 100,000. The CDC is now claiming: overall 178 per 100,000 as of 2007.

The Organization for Economic Cooperation and Development (OECD) Health Data 2010

Could this be because the CDC information is governed by the drug companies?

Why does anybody know this?

It can take hours or even days to find raw data on cancer and its history—you may have to pour hours of time to find the correct information. The painful truth is that this data is purposely hidden by overcomplicating it. The data that is available to everyone is has one thing in common more people are dying of cancer now than ever before!

All the media and journal articles try to contradict that fact and claim that cancer rates are falling, but when you really look at food and the chemicals that are in it, and look at the cancer mortality rates, you will clearly see that it is simply not true. More

38

people are becoming diagnosed with different types of cancer every day! Unfortunately, the overall population is oblivious to the fact that we are creating the war on cancer!

Even though cancer has been around for thousands of years and might, in fact, have been around from the beginning of life on this earth, we have been experiencing a leap in numbers. Cancer used to be uncommon a hundreds of years ago, and cancer was virtually unheard of in the USA.

In the not-too-distant past, people relied on whole and unrefined foods for their nutrients, definitely not genetically modified. After World War I, the processed food revolution began, and proportions became a greater part of an unhealthy diet in the everyday life of the average American home. It started with the canning industry, later became the food processing industry. Shortly after the boom of the processing revolution came pasteurization, chemical additives, bleaching, and other procedures which are now part of mass food production. Food processing was to make the life of food on the shelves last as long as possible in hopes of increasing overall profits and allowing more access to food for more people.

This was accomplished by getting rid of natural enzymes from the food through food processing. As the result, food has lost its nutritional value, resulting in foods that have almost no health benefits and could cause malnutrition over a period of time. Technology for more dependable methods of taking these important enzymes out of the food eventually became better, and the shelf life increased. Unfortunately, the health value of the food decreased right along with it.

Section 6: Part: 2
Oxygen to Fight Cancer

In the 1920s, a doctor and Nobel Prize winner, Otto Warburg, and Dr. Emil Fischer were studying in the field of the polypeptide. What is a polypeptide. You might ask? This is a polymer of amino acids linked together by peptide bonds. Descriptive, right? It's essentially a protein. Their research was specifically in the field of oxidation. This research led the investigation into the physical and chemical methods, eventually leading to detailed studies on the integration of carbon dioxide in plants and the metabolism of tumors. Warburg's latest research at the Kaiser Wilhelm Institute led to the discovery that flavins and nicotinamide are the active means of the hydrogen-transferring enzymes.

Fermentation is very conducive to cancer research; this is only referring to partially processed carbohydrates. Everyone knows that sugar ferments, and the result of such a process is a release of gases. These gases replace oxygen in normal body cells. A low-oxygen environment is perfect for cancer growth! In earlier research by Otto Warburg, cancer can't live in highly oxygenated environments, but it thrives on sugar. Understanding this one simple thing is paramount! However, it is almost impossible to escape sugar in foods these days. **"All normal cells have an absolute requirement for oxygen, but cancer cells can live without oxygen—a rule without exception."**… "Deprive a cell 35% of its oxygen for 48 hours and it may become cancerous." Dr. Otto Warburg, the Kaiser Wilhelm Institute for Cell Physiology.

From *Thomas Jefferson University and Cancer Biology & Therapy Journal*, a study conducted by researchers suggest that antioxidants can help in the fight against cancer. Scientists have known from Otto Warburg's research that diets high in antioxidant-containing foods can not only help fight cancer, but also cure cancer and help prevent it altogether. **"Cancerous tissues are acidic, whereas healthy tissues are alkaline. Water splits** into H+ and OH- ions. If there is an excess of H+, it is acidic; if there is an excess of OH-ions, then it is alkaline." Dr. Otto Warburg, the Kaiser Wilhelm Institute for Cell Physiology

"Reactive oxygen species [free radicals] are formed in vivo during normal aerobic metabolism and can cause damage to DNA, proteins, and lipids, despite the natural antioxidant defense system of all organisms," writes Erich Grotewold in the book *The Science of Flavonoids*. "Diets high in flavonoids, fruits, and vegetables are protective against a variety of diseases, particularly cardiovascular disease and some types of cancer."

The Thomas Jefferson University researchers published on 15 May, 2013, that they had discovered a protein called Caveolin-1 (Cav-1) that when present in the human genome can suppress tumor growth. The existence or nonexistence of the Cav-1 protein is the best way of foretelling the end result of breast cancer. In women with triple-negative breast cancer, a very aggressive disease with no cure, 75% percent are more likely to be alive twelve years after diagnosis if they have the Caveolin-1 protein, and less than 10% are likely to be alive for five years after diagnosis if Caveolin-1 is nonexistent. The investigators further verified that Cav-1 plays an antioxidant role in the body. In the current study, the scientists confirmed that when Cav-1 is removed, oxidative stress in breast cancer tumors increases, which is a huge 300% growth in tumor mass and volume. "Antioxidants have been associated with cancer-reducing effects—beta carotene, for example—but the mechanisms, the genetic evidence, have been lacking," lead researcher Michael P. Lisanti said. "This study provides the necessary genetic evidence that reducing oxidative stress in the body will decrease tumor growth."

The findings of cancer and oxygen contradict all of the information that most people are aware of. Many people don't know that the effectiveness of chemotherapy drugs is no more effective than a placebo pill. Chemotherapy drugs are well known to lower the oxidation within the body, and some research indicates that many antioxidant drugs used to treat other disorders, such as diabetes and malaria, may be effective in treating cancer.

Section 7: Part: 1
Iodine in the Fight for your Health

Nascent iodine is great to help prevent cancer by cleaning the body's glands and organs from the buildup of toxins that can result in lower oxygen levels in the blood. There are many kinds of iodine—some that the body can use, and others that will only make expensive urine. Potassium iodine is one of the most common used by many people around the world. The trouble is that there are many false and useless forms of potassium iodine.

Iodine is a chemical component that's essential for the healthy function of the human body, however, the body can't make iodine. There are many glands and organs in the body that require iodine, the most commonly known is the thyroid gland. The thyroid gland uses the iodine to balance the hormones it creates if you do not receive the adequate quantity of iodine in your diet this will offset the hormones in your body.

Iodine deficiency is commonly contributed to an enlargement of the thyroid gland (goiter). But not so widely known are the other serious effects that iodine deficiency can cause, low thyroid hormone can induce a woman to stop ovulating, leading to infertility. Research indicates that iodine deficiency is linked to cancers such as "prostate, breast, endometrial, and ovarian cancer."

Iodine is so crucial to your health that it aids in fighting off disease, there have been many successful treatments using iodine. Such treatments include skin "fungus, fibrocystic breast disease; preventing breast cancer, eye disease, diabetes, and heart disease and even stroke."

Iodine is a vital trace element required for the synthesis of hormones, and the absence of it can also cause or contribute to the development of:

• Hypothyroidism (in which the thyroid gland doesn't make adequate thyroid hormone)

43

• Goiter (a swelling of the neck or larynx resulting from enlargement of the thyroid gland)

• Mental retardation

• Cretinism (severely stunted physical and mental growth and deafness due to untreated congenital hypothyroidism)

• Certain forms of cancer

It's also important to realize that there are two types of iodine, and that you need them both:

1. Iodine

2. Iodide (eating a healthy, balanced diet usually supplies your daily requirements)

One of the prime differences between iodine and iodide, is that iodine will diffuse into the cells while iodide must be transported into each cell. All tissues in your body are different, some will absorb iodine while others will absorb iodide, you need both supplements for optimal health.

Iodide/Iodate and the Health Risk

The truth about iodide and iodate is really hard to find, and can't be taken at face value. Compounds with iodine in the formal oxidation state are called iodides, and potassium iodate is an oxidizing agent. Potassium iodate is sometimes used for iodination of table salt to prevent iodine deficiency. Because iodide can be oxidized by molecular oxygen to iodine under wet conditions, U.S. companies add thiosulfates or other antioxidants to the potassium iodide. In other countries, potassium iodate is used as a source of dietary iodine. It is also an ingredient in some baby formulas.

Several Internet-based companies have been marketing potassium iodate (KIO3) for radiation protection in place of potassium iodide (KI). Though the names are similar, the products are very different, and the U.S, Food and Drug Administration have expressed serious concerns about the safety and effectiveness of iodate, and the fact that its manufacturers are not in conformity with FDA rules to assure safety, quality, and purity of the product. Although the FDA has been successful at removing most iodate products from store shelves, iodate can still be found on the Internet, and is falsely claimed to be FDA approved on Wikipedia. Wikipedia unwittingly allows dosage charts and other false claims concerning this unapproved drug to be posted by sellers of iodate who edit the page.

In the MSDS for Potassium Iodate (KIO3), it stated:

"Potential Health Effects

Eye: May cause eye irritation. May cause conjunctivitis. May cause permanent corneal opacification.

Skin: May cause severe irritation and possible burns.

Ingestion: May cause burns to the gastrointestinal tract. May cause nausea, vomiting, and diarrhea, possibly with blood.

Inhalation: May cause acute pulmonary edema, asphyxia, chemical pneumonitis, and upper airway obstruction caused by edema.

Chronic: Prolonged or repeated skin contact may cause irritation. Prolonged or repeated exposure may cause gastrointestinal irritation and kidney damage. Chronic ingestion may cause central nervous system failure. Effects may be delayed."

Would you take a pill that claims it can make you healthy, but also can cause long-term health problems? The FDA then approves the use of potassium iodide (KI) within the United States of America over the past few years. It is harder and harder not only to get the right information on potassium iodide (KI), but also to get your hands on the right kind of potassium iodide.

Stay away from potassium iodide products that are not FDA approved.

ThyroSafe (KI): **"Protects the thyroid from absorbing cancer-causing radioactive iodine released from a nuclear reactor or nuclear bomb. "Protects the thyroid from absorbing cancer-causing radioactive iodine released from a nuclear reactor or nuclear bomb.**

ThyroSafe is the only FDA-approved potassium iodide (KI) tablet with 65 mg. This is essential for promptly dosing children, who are always more at risk than adults. The FDA Guideline suggests: **"For the sake of logistical simplicity in the dispensing and administration of KI to children, FDA recommends a 65-mg. Dose as standard for all school-age children while allowing for the adult dose (130 mg. 2 X 65 mg. tablets) in adolescents approaching adult size." Read instructions thoroughly prior to ingesting."**

Potassium iodate (KIO3) is not only unhealthy and unsafe, but can cause long-term damage to the body, whereas potassium iodide (KI)

(I recommend the ThyroSafe) is not only good for you, but shouldn't be taken unless a nuclear event (such as Fukushima) has occurred with evidence that the radioactive particles can or will be in an area near you.

Section 7: Part: 3
What are the Benefits of Iodine?

Iodine deficiency can start in childhood and can damage your child's IQ. Iodine is used by all of your glands and uses by the mucus membranes in your body. Iodine helps fight cancer, and the escalation of gland-related cancers are on the climb, as well as cystic diseases. In some cases, doctors have been found to tell their patients that they don't need iodine, which is just not true! Iodine is also used by the **adrenal glands, parotid glands, mucosal lining of the intestines, pineal gland, pituitary gland, thyroid gland, thymus, pancreas, salivary glands (mouth), breathing system (nose, sinuses, lungs).**

Urinary system: (kidneys, ureter, bladder, urethra), sexual organs (breasts, ovaries, vagina, uterus, prostate, testes), liver and gallbladder (bile), brain (substantia nigra, choroid plexus), eyes (conjunctiva, ciliary), skin, bone marrow, and many, many more.

Many Asian cultures such as the Japanese have lower cancer rates than any other country in the world. This is largely due to their iodine-rich diet. I have heard of people telling me that they are allergic to iodine, but are they really allergic? An allergy to iodine is tremendously uncommon. Having an allergy to iodine is very common to mistake. Some people think that if they are allergic to seafood, they are allergic to the iodine contained within the seafood. This is not true. Iodine is found in all the glands listed above. Without iodine in our bodies, there can't be any human life. Both freshwater and saltwater fish contain iodine, and there are many foods that have small amounts of iodine in them, including yogurt, milk, bread, macaroni, eggs, corn, and many beans. Allergic reactions to shellfish and fish have been linked to the proteins within the muscles (tropomyosins, parvalbumin). The allergy has nothing to do with iodine.

The effects that many people associate with an allergic reaction can include, but are not limited to: acne, rash, headaches, heart palpitations, aching joints, nasal congestion, and postnasal drip, this is a sign of weak adrenals. In some cases, when supplementing with a thyroid hormone, you may need to lower your dosage. Please, consult your doctor before taking iodine. Iodine is a cleaning and detox agent, and many people are not prepared for the purging of the toxins that takes place, and might think that iodine is causing the problem.

The truth, in fact, is the opposite—it's the body ridding itself of the harmful toxins. I would recommend learning how to identify the symptoms and manage the iodine, according to your detox symptoms. I would suggest cutting down the dose, or just pushing through and keeping on the same dose, or even increasing it, this would rid the body of toxins faster instead the prolonging uncomfortable side effects.

Mild iodine deficiency can include weight gain, sluggishness, fatigue, fuzzy brain, and cold extremities. After I did my research on iodine and foods that contain the iodine naturally, I learned of centuries'-old knowledge with many studies that show iodine not only fights off cancer, but can even help prevent cancer.

I wondered if the media are really telling you the truth about how to fight cancer. My findings, there are two principal components in fighting and preventing cancer. Number one is increasing the oxygen in the blood and body, and number two is using iodine for removing the toxins that can cause cancer. *7 Foods Rich in Iodine*, published on globalhealingcenter.com on January 18, 2011, has a video with Dr. Group, one of my favorite doctors, who speaks out about the power of iodine, and advocate of some of the highest-quality iodine on the market (Survival Shield x2). **"Iodine therapy helps the body eliminate fluoride, bromine, lead, cadmium, arsenic, aluminum and mercury. The Recommended Daily Allowance (RDA) for iodine is 150 micrograms daily for everyone over the age of 14. The RDA for children ages 1-8 is 90/mcg every day, ages 9-13 is 120/mcg every day. And if you're pregnant or breastfeeding, it is recommended that you get 290/mcg every day."**

Iodine Rich Foods

1. **Organic Sea Vegetables**

The life in the oceans is abundant and rich in iodine, kelp has the most iodine of any other food! Just one tablespoon of kelp has 2000/mcg of iodine, other sea veggies are arame contains 730/mcg, Hiziki contains 780/mcg of iodine, kombu contains 1450/mcg of iodine, wakame contains 80/mcg of iodine.

2. **Organic Cranberries**

Cranberries are great for thanksgiving, they are also full of antioxidants, but did you know that they are also a great source of iodine? Only four ounces of cranberries has 400/mcg of iodine. Fresh cranberry juice will also contain iodine, make sure the juice is organic and has the least amount of ingredients, and low sugar contents.

3. **Organic Yogurt Organic**

Greek yogurt is a great addition to your diet, only one cup of Greek yogurt has about 90/mcg of iodine.

4. **Organic Navy Beans**

Everyone knows the health benefits of beans, great source of fiber, and protein. Navy beans are also loaded with iodine, just a 1/2 a cup of Navy beans has 32/mcg of iodine.

5. **Organic Strawberries**

This delightful fruit contains 13/mcg of iodine for just one medium strawberry.

6. **Himalayan Crystal Salt**

Himalayan salt is also known as gray salt, and pink salt, unlike table salt this natural salt maintains all of their nutritional value, table salt has been stripped of the vital minerals. Table salt has been chemically cleaned and is made up of mostly sodium chloride. Some table salts are also made from or crude oil flakes cooked at 1200° Fahrenheit, the danger of table salt does not end with what it may be made with. The processing of table salt may contain some very toxic chemicals such as Sodium aluminosilicate, fluoride, anti-caking agents, toxic amounts of aluminum derivatives. Sea salts such as Himalayan salt are natural, I recommend organic sea salt just one gram of the sea salts contain 500/mcg of iodine along with many other health benefits.

7. Organic Potatoes

Potatoes have been a staple food in many cultures around the world, this great veggie is also loaded with iodine, the common potato has more iodine than any other veggie. Just one medium-sized potato has 60/mcg of iodine and the skin contains the most iodine in all the potato. If iodine is in so many kinds of plants and animal-based foods, why does the United States have such high cancer rates?

The variety of iodine deficiency disorders:

Fetus	Miscarriage Stillbirths Congenital anomalies Increased perinatal morbidity and mortality Endemic cretinism
Neonate	Neonatal goiter Neonatal hypothyroidism Endemic mental retardation Increased susceptibility of the thyroid gland to nuclear radiation
Child and adolescent	Goiter(subclinical) hypothyroidism Impaired mental function Retarded physical development Increased susceptibility of the thyroid gland to nuclear radiation
!Adult	Goiter with its complications Hypothyroidism Impaired mental function Spontaneous hyperthyroidism in the elderly Iodine-induced hyperthyroidism Increased susceptibility of the thyroid gland to nuclear radiation

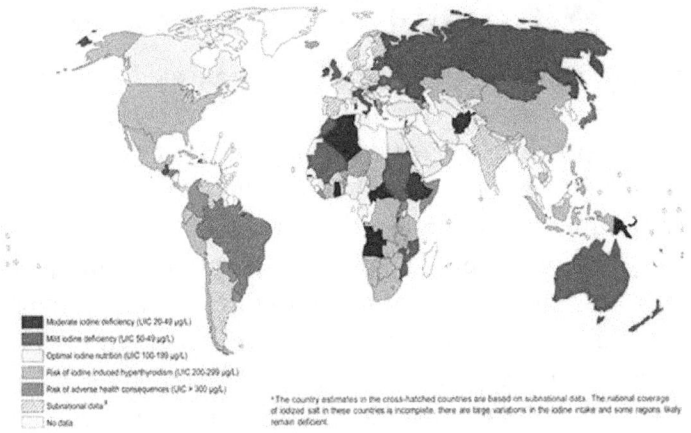

Source: thyroidmanager.org

Iodine levels in people in the USA have fallen because environmental toxins, such as bromine, fluoride, and chlorine can prevent its absorption within the human body. Bromine is used in making flame retardants fabrics, carpets, new cars, pesticides, and drugs, and is found in many baked products and processed foods. About 74 years ago, the food industry removed iodine from baked goods and replace iodine with bromine.

Chlorine is found in water, as well as toxic compounds like PCBs (polychlorinated biphenyls). Fluoride is often added to toothpaste, water, and is found in some drugs such as Prozac and SSRIs. Iodine helps the body rid itself of harmful chemicals such as mercury, bromine, and fluoride.

Section 7: Part: 4
Why Iodine Over Bromine?

    Sometime in the 1920s, the FDA requested that bakers add iodine to bread in substantial quantities. This was meant to help—and did help—prevent iodine deficiency and goiter. Back in those days, a single slice of bread would contain 150 mg., which would fill the minimum recommended dietary allowance (RDA) for the full day. In 1960, the average daily American diet contained full RDA of 0.15 mg iodine., with baked products being responsible, for about to 75 % of total iodine intake, which served to prevent underactive thyroid symptoms. Around 74 years ago,
the food manufacturers, for some reason, made a choice to remove iodine from baked products and exchanged it for something toxic—bromine.

    Bromine in the body acts as a Trojan horse and mimics iodine, appearing similar to the thyroid gland. It tends to bind itself very easily, bromine doesn't benefit the thyroid gland in any way and actually prevents the absorption of iodine. Bromine is also linked directly to irritability, dizziness, drowsiness, and impaired thinking and memory. Today's "iodized" salt is useless, as it contains very high levels of bromine rather than iodine. This substitution of bromine has resulted in nearly universal iodine deficiency in the American population.

JULY 19, 1999 12:00 Noon

FOR IMMEDIATE RELEASE

CONTACT: Center for Science in the Public Interest 202/332-9110

**Consumer Group Calls for Ban on "Flour Improver";
Potassium Bromate Termed a Cancer Threat**

    WASHINGTON - July 19 - The Center for Science in the Public Interest today petitioned the Food and Drug Administration (FDA) to prohibit the use of potassium

bromate, which is used to strengthen bread dough. CSPI charged that the FDA has known for years that bromate causes cancers in laboratory animals, but has failed to ban it. Bromate was first found to cause tumors in rats in 1982. Subsequent studies on rats and mice confirmed that it causes tumors of the kidney, thyroid, and other organs. Instead of banning bromate, since 1991 the FDA — with only partial success — has urged bakers to voluntarily stop using it.

"The FDA should fulfill its responsibility to protect the public's health," said Michael F. Jacobson, Ph.D., Executive director of CSPI. "Instead of meeting privately with industry, the FDA should ban bromate immediately."

"In 1992-93 and again in 1998-99, the FDA tested several dozen baked goods and found that many contained bromate at levels considered unsafe by the agency," said Darren Mitchell, a CSPI attorney. "One sample tested recently had almost 1,000 times the detection limit. The FDA's inaction needlessly exposes consumers to this harmful additive."

Food additives that cause cancer usually can be banned under the Delaney clause of the Food, Drug, and Cosmetic Act. However, because the FDA sanctioned the use of bromate before the Delaney clause went into effect in 1958, it is harder for the agency to ban the substance. Bromates have been banned in numerous countries, including the United Kingdom in 1990 and Canada in 1994. In addition, in 1991, California declared bromate a carcinogen under the state's Proposition 65. Baked goods sold in California would have to bear a cancer warning if they contained more than a certain level of bromate. As a result, most California bakers have switched to bromate-free processes.

Many bakers, including Best Foods, Inc. (Maker of Arnold, Entenmann's, and Oroweat brand breads and rolls), Pepperidge Farm, and Pillsbury, have switched to bromate-free processes. Also, some supermarket chains, including Giant, Jewel, Ralph's, and Von's, do not use bromate. In contrast, Interstate Brands Corp. (Wonder, Home Pride), Schmidt Baking Co. (Schmidt, Sunbeam), Tasty Baking Co. (TastyKake), and Martin's still use potassium bromate in some of their products. Among fast-food chains, Burger King, Arby's, and Wendy's use bromate in buns, and Boston Market uses it in its French sandwich bread. CSPI advises consumers to avoid bread, rolls, doughnuts, and cakes that list 'potassium bromate' or 'bromated flour' among their ingredients"

The FDA's limited surveys found that rolls and buns are especially likely to contain high levels of bromate. As iodine levels fell over the last 40+ years in the United States, autoimmune disorders and hyperthyroid symptoms have been increasing at near epidemic proportions. As a direct effect of the increased levels of fluorine, bromine, and chlorine (toxic halides) in the environment and in the food supply, iodine levels have not only fallen, but larger quantities of iodine are needed to correct iodine deficiency as well as to promote a detoxifying effect of hard toxic metals.

Iodine consumption immediately increases the detoxification of bromine, fluoride, and some heavy metals including mercury, lead, cadmium, and aluminum. Bromine and fluoride are only removed by iodine, and no other detoxifying technique. Many people know that toxic halides, such as fluoride and bromine, have an effect on the receptors in the body similar to iodine, and can inhibit iodine absorption and binding in the body. Even though iodine is part of the halide family, which are all opposing elements to iodine, they will obstruct the absorption of iodine. We are all exposed to bromine and fluorine, your bodies are storing these toxic chemicals in the receptors meant for iodine. The human body's receptors are primitive and lacking in sophistication, so the cells are not able to differentiate iodine from other anions of similar atomic or molecular size, which may act as "pseudo-iodides": bromine, fluoride, chlorine, thiocyanate, cyanate, nitrate, Pertechnetate, and perchlorate. As nature had intended, for everything that can hurt the human body, there is something to reverse the damage, iodine is the only option when it comes to removing these toxic halogens from the thyroid and even the pineal gland, where fluoride concentrates.

Humanity is ultimately leading their children down a path of doom—every child who is born in the USA is being poisoned. **"overwhelming evidence that every child, no matter where in the world he or she is born, will be exposed, not only from birth, but from conception, to man-made chemicals that can undermine the child's ability to reach his or her fullest potential—chemicals that interfere with the natural chemicals that tell tissues how to develop and construct healthy, whole individuals according to the genes they inherited from their mothers and fathers,"** wrote Dr. Theo Colborn, Senior Program Scientist, at the World Wildlife Fund.

Section 7: Part: 5
What Foods Inhibit the Absorption of Iodine

Even with a healthy, balanced diet and adequate iodine intake, there are some foods that can lessen the absorption of iodine. Before we start the section off, let's have a little review. Here is a short list of foods that are rich in iodine:

World's Healthiest Foods rich in iodine		
Food	Cals	%Daily Value
Sea Vegetables	9	276.6%
Yogurt	154	58.1%
Milk	122	39%
Eggs	78	18%
Strawberries	46	8.6%
Cheese	72	6.7%

Because the human body is a balanced system, there is one thing that will lessen its health benefits. In this section, we'll have a look at the foods that may inhibit the body from getting the iodine it needs. There are many reasons why these foods prevent your body from getting the amounts of iodine you need. I'm not telling anyone not to eat any of the foods in this section—many of them have great nutritional value. Through a balanced diet, your body can prevent cancer and help treat or even fight off other medical conditions, and you'll have more energy and overall health.

Goitrogens are natural chemicals that will suppress the natural thyroid function. They interrupt the glands that require iodine, and in effect, block them from getting the iodine. Some of the elements in goitrogens encourage the antibodies that react to the thyroid gland, and some of the chemicals restrict the thyroid peroxidase (TPO), the mechanism responsible for the accumulation of iodine for making thyroid hormones.

Here's a list of symptoms that low levels of iodine can cause:

- Severe fatigue, loss of energy
- Weight gain, difficulty losing weight
- Depression and depressed mood
- Joint and muscle pain, headaches
- Dry skin, brittle nails
- Brittle hair, itchy scalp, hair loss
- Irregular periods, PMS symptoms
- Lower levels of Breast milk
- Calcium metabolism difficulties
- Difficulty tolerating cold and lower body temperature
- Constipation
- Sleeping more than average
- Diminished sex drive
- Puffiness in face and extremities
- Hoarseness
- Bruising/clotting problems
- Elevated levels of LDL (the "bad" cholesterol) and heightened risk of heart disease
- Allergies that suddenly appear or get worse
- Persistent cold sores, boils, or breakouts
- Tingling sensation in wrists and hands that mimics carpal tunnel syndrome
- Memory loss, fuzzy thinking, difficulty following conversation or train of thought

• Slowness or slurring of speech

These symptoms are also associated with hypothyroidism, and many of them can be completely fixed with the right intake of iodine and a healthy diet.

What foods contain goitrogens?

Here are some of the foods I would completely cross off your grocery list and cut out of your diet. These foods will be discussed later in the book or have been already.

Gluten:

For many people, gluten wouldn't come to mind when we talk about foods with goitrogens, but gluten is not only linked to celiac disease. Gluten sensitivity has been found to go hand-in-hand with autoimmune disorders, a very aggressive form of allergy, such as type 1 diabetes, Addison's disease, Sjögren's syndrome, rheumatoid arthritis, and autoimmune thyroid disease. Many doctors would tell their patients to consider excluding gluten from their diets, especially if they already have an autoimmune disorder.

Soy:

Soy is a food that contains isoflavones (estrogen mimickers) and it also contains goitrogenic compounds. These compounds, just like thyroid hormones, accept iodine molecules from the thyroid peroxidase (TPO). Some of the research studies state that genistein and isoflavones may block the thyroid hormones for iodine and block the production of TPO.

Isothiocyanates:

These compounds are primarily found in cruciferous vegetables, such as cabbage, Brussels sprouts, broccoli, broccolini, cauliflower, mustard greens, kale, turnips, and collards. Isothiocyanates seem to block Thyroid peroxidase (TPO). Some research suggests that they may also interrupt communication across the thyroid's cell membranes. I'm not suggesting that these vegetables are bad for you—they are chock full of vitamins, minerals, antioxidants, and a variety of nutrients we all need. People with thyroid problems definitely shouldn't avoid them, just increase your iodine.

There's a very good trick to see how iodine deficient you are. These foods would taste bitter to a person who's deficient in iodine, and there is a biological reason for this. Your body is telling you that it needs something, and that's not the right thing. As the body is lacking in iodine, the foods that are rich in iodine would taste good, and the foods that will block or inhibit the iodine would taste bitter.

- Bok choy
- Broccoli
- Brussels Sprouts
- Cabbage
- Cauliflower
- Kale
- Millet
- Mustard seeds
- Peanuts
- Pine nuts
- Rutabagas
- Spinach
- Turnips

There are chemicals that are known to block iodine absorption. Some of the chemicals have been discussed in parts of this book or will be talked about in more detail.

- Refined white sugar is mostly made with high fructose corn syrup and beet sugar. This will make the thyroid burn out over time, and not function properly.

- Fluoride

- Chlorine

- Bisphenol, also known as BBA.

- Mercury confuses the receptors that receive the iodine and blocks the iodine as it moves into place.

- Gluten increases thyroid antibody levels.

- Soybeans, soy oil, soy proteins

- Pasteurized milk—grain-fed cows fed on feedlots lack iodine. Pasteurization kills iodine. Antibiotics and growth hormones block iodine. Synthetic and inorganic vitamins are added to pasteurized milk, which can cause a whole multitude of troubles.

Section 8: Part: 1
What are the Types of Fluoride?

When fluorine is combined with something, it becomes a fluoride compound. Fluoride range is very large, due to fluorine-containing combinations, and it has the ability of mutating with nearly all the elements.

Here are some common kinds:

Sodium fluoride is a colorless, crystalline, water-soluble, poisonous, solid, NaF, used primarily in the fluoridation of water, as an insecticide, and as a rodenticide (a substance or preparation for killing rodents).

Sodium fluoride is used in most toothpastes, mouthwashes, dental varnish, dental preparations, and nutritional supplements. Sodium fluoride is also utilized in the fabrication of chemical and biological weapons.

Calcium fluoride (CaF2) is compounded of calcium and fluorine, which occurs naturally as the mineral fluorite also called fluorspar. Most of the world's fluorine comes from calcium fluoride. Fluorides, in general, are toxic to humans. However, CaF2 is considered the least toxic, and even relatively harmless due to its ability to dissolve in liquid. Calcium fluoride is a well-known remedy for fluoride poisoning. When a cure exists in a mixture with a poison, it makes the poison far less toxic to the body. Calcium fluoride is the form of fluoride normally found in natural, untreated waters.

Cryolite is used as pesticide often applied directly to field crops, resulting in creating fluoride residues in and on fresh fruits and vegetables. The main way people are exposed to fluoride from the pesticide cryolite is through ingesting of grape products, mostly white grapes, grown in the U.S. According to data from the USDA in 2005, the average fluoride levels in grape products are:

• White grape juice = 2.13 ppm

- White wine = 2.02 ppm

- Red wine = 1.05 ppm

- Raisins = 2.34 ppm

Many juice drinks that are not labeled as "grape juice" use grape juice as a filler ingredient. The use of cryolite thus contaminates many juices with fluoride. As of 2011, the EPA is calling to increase the allotted amounts of cryolite that's in the food.

EPA's new March 2011 proposal for cryolite residues on foods:

Commodity	Current	Proposed
cabbage	7 ppm	45 ppm
citrus fruits	7 ppm	95 ppm
collards	7 ppm	35 ppm
eggplant	7 ppm	30 ppm
lettuce	7 ppm	
lettuce, head		180 ppm
lettuce, leaf		40 ppm
peaches	7 ppm	10 ppm
raisins	None	55 ppm
tomatoes	7 ppm	30 ppm
tomato paste	None	45 ppm

Fluorosilicic acid (H_2SiF_6) is most commonly used for water fluoridation, also known as hydrofluorosilicic acid. This form of fluoride is a toxic liquid byproduct, acquired by scrubbing the chimney stacks of phosphate fertilizer manufacturers. Other names for it are hexafluorosilicic, hexafluorosilicic, hydrofluosilicic, and silicofluoric acid. The CDC approximates that 95% of our water is fluoridated with fluorosilicic acid. Fluorosilicic acid is a waste product of the phosphate fertilizer industry, and is heavily contaminated with toxins and hard metals (including the carcinogens arsenic, lead, and cadmium) and radioactive materials.

This material is the waste residue from the superphosphate fertilizer industry, and about 70-75% percent of this stuff comes from the Cargill manufacturing company out of Minneapolis, MN.

Sodium (Na_2SiF_6) is mainly added to public drinking water as a fluoridation agent. This is also used as an insecticide. It's a classified hazardous waste byproduct of phosphate fertilizer manufacturing, which, if not put into our drinking water, must be disposed of at hazardous waste facilities. Na_2SiF_6 is also known as sodium fluosilicate and sodium silica fluoride.

Stannous fluoride is the popular name given to fluoride. Stannous fluoride is an additive in Crest Pro-Health toothpaste because it doesn't become organically inactive in the presence of calcium, as sodium fluoride does.

Sulfuryl fluoride leaves two different residues on food— fluoride and sulfuryl fluoride. The EPA has given legal acceptances for each. To this day, we know very little about the chemical Sulfuryl fluoride itself, apart from the fact that it is extremely toxic and attacks the brain and nervous system. The use of sulfuryl fluoride on or around food products was not allowed due to its toxicity. In 2004, the EPA reversed this policy, pushed by the companies that make it, resulting in the EPA allowing its use on food. Processing companies nationwide cleanse their facilities with sulfuryl fluoride, leaving behind high levels of fluoride in and on foods and even food packaging. Sulfuryl fluoride cleansings produce fluoride residues of 70 ppm on processed foods and 130 ppm on wheat. There have been no labeling requirements for foods treated with sulfuryl fluoride. Consumers have no way of knowing which food companies are using it. In January of 2011, the right to use sulfuryl fluoride was reversed and then stopped.

Section 8: Part: 2
Detox Fluorides

Cleansing your body from the toxic chemical fluoride, you can remove fluoride with some easy natural remedies. Fluoride has been related to an assortment of severe and chronic health problems. Fluoride is an artificial waste product of nuclear, aluminum, and phosphate fertilizer companies. Fluoride has the ability to combine and increase the toxicity of other toxic materials. Fluoride acquired from industrial waste and added to water supplies it is also polluted with lead, aluminum, and cadmium. The fluoride used for water fluoridation doesn't have FDA approval and is viewed by the FDA as an "unapproved drug." Any drug requires clarity on the limits that the human body can ingest. Fluoride is in your food, drinks, and dental hygiene products. Research shows that fluoridation is redundant, since we're already getting 300% of the American Dental Association's recommended daily amount.

From a 1951 American Dental Association brochure: **"There is no proof that commercial preparations such as tablets, dentifrices, mouthwashes or chewing gum containing fluorides are effective in preventing dental decay. Unfortunately, such preparations are being offered to the public without adequate scientific evidence of their value."**

People who live in fluoridated areas are having effects ranging from:

> • **Risk to the brain**. According to the National Research Council (NRC), fluoride can damage the brain. Animal studies conducted in the 1990s by EPA scientists found dementia-like effects at the same concentration (1 ppm) used to fluoridate water, while human studies have found adverse effects on IQ at levels as low as 0.9 ppm among children with nutrient deficiencies, and 1.8 ppm among children with adequate nutrient intake.

65

• **Risk to the thyroid gland**. According to the NRC, fluoride is an "endocrine disrupter." The NRC has warned that doses of fluoride (0.01-0.03 mg/kg/day), achievable by drinking fluoridated water, may reduce the function of the thyroid among individuals with low iodine intake. Reduction of thyroid activity can lead to loss of mental acuity, depression, and weight gain.

• **Risk to bones.** According to the NRC, fluoride can diminish bone strength and increase the risk of bone fracture. While the NRC was unable to determine what level of fluoride is safe for bones, it noted that the best available information suggests that fracture risk may be increased at levels as low 1.5 ppm, which is only slightly higher than the concentration (0.7-1.2 ppm) added to water for fluoridation.

• **Risk of bone cancer**. Animal and human studies—including a recent study from a team of Harvard scientists—have found a connection between fluoride and a serious form of bone cancer (osteosarcoma) in males under the age of twenty. This connection has been described by the National Toxicology Program as "biologically plausible." Up to half of adolescents who develop osteosarcoma die within a few years of diagnosis.

• **Risk to kidney**. People with kidney disease have a heightened susceptibility to fluoride toxicity which stems from an impaired ability to excrete fluoride from the body. As a result, toxic levels of fluoride can accumulate in the bones, intensify the toxicity of aluminum build-up, and cause or exacerbate a painful bone disease known as renal osteodystrophy.

"I am appalled at the prospect of using water as a vehicle for drugs. Fluoride is a corrosive poison that will produce serious effects on a long-range basis. Any attempt to use water this way is deplorable."

- Dr. Charles Gordon Heyd, Past President of the American Medical Association.

"I would advise against fluoridation. Side effects cannot be excluded. In Sweden, the emphasis nowadays is to keep the environment as clean as possible with regard to pharmacologically active and, thus, potentially toxic substances."- Dr. Arvid Carlsson, co-winner of the Nobel Prize for Medicine (2000).

Fluoride not only damages the liver and kidneys, but it also is known to weaken the immune system, which can lead to cancer. According to the National Toxicology Program (NTP), "the majority of evidence" from laboratory studies specifies that fluoride is a mutagen (a compound that can cause genetic change).

A chemical that can cause genetic change is one that can likely cause, or contribute to, the progress of cancer. Fluoride can even mimic systems of other conditions such as fibromyalgia (a syndrome in which a person has long-term, body-wide pain and tenderness in the joints), it will also carry aluminum across the blood brain barrier. This in effect causes notorious "dumbing down" with lower IQs and Alzheimer's effects of fluoride.

One paper entitled *Fluoride—A Modern Toxic Waste* says the following: **"Fluoride increases the tumor growth rate by 25% at only 1 ppm, produces melanotic tumors, transforms normal cells into cancer cells and increases the carcinogenesis of other chemicals."** For the original references to these studies, refer to Yiamouyiannis' pamphlet, Lifesavers Guide to Fluoridation.

In 1977, epidemiological studies by Dr. Dean Burk, former head of the Cytochemistry Section of the National Cancer Institute and Dr. John Yiamouyiannis shown that fluoridation caused about 10,000 cancer deaths over a 15-year period. Even though the findings occurred in 1977, they were released in 1989. After researchers analyzed the results in rats, it was later discovered that animals who drank fluoridated water:

• Showed an increase in tumors and cancers in oral squamous cells.

• Developed once was a rare variety of bone cancer, now is considered to be a common type of cancer called osteosarcoma.

• Showed an increased in thyroid estrous cycle cell abnormal growth of tissue in some part of the body.

• Developed a rare form of liver cancer known as hepatocholangiocarcinoma (HCC-CC).

It was not until the 1990s that research shown the impact and effects of fluoride on the pineal gland, which is located between the two hemispheres of the brain. It's responsible for regulating the production of the hormone melatonin, which regulates the start of puberty. The pineal gland protects the body from cell damage caused by free radicals and helps you stay younger, according to research done by Dr. Jennifer Luke from the University of Surrey in England. The pineal gland is the primary target of fluoride buildup within the body.

The tissue of the pineal gland contains more fluoride than any other soft tissue or gland in the human body. Dr. Luke summarized their findings on fluoride buildup in humans: **"In conclusion, the human pineal gland contains the highest concentration of fluoride in the body. Fluoride is associated with depressed pineal melatonin synthesis by prepubertal (the early onset of puberty in females) desert rats and an accelerated onset of sexual maturation in female desert rats. The result strengthens the hypothesis that the pineal gland has a role in the timing of the onset of puberty. Whether or not it interferes with the final function in humans requires further investigation."**

Research shows that the pineal gland is not only responsible for sleep-wake patterns, aging, puberty, maturity, and IQ, fluoride contributes to thyroid problems that affect the entire endocrine system. Abundant variations of fluoride are also in many insecticides for homes, and pesticides for crops. There is even fluoride in baby food and bottled waters. By purchasing commercially grown fruits, especially grapes, and vegetables that are chemically sprayed and grown in areas irrigated by fluoridated water, you're slowly killing your endocrine system, and damaging your and your children's IQ.

The following letter was received by the Lee Foundation for Nutritional Research, Milwaukee, Wisconsin, on 2nd October 1954, from a research chemist by the name of Charles Perkins.

Charles writes: **"I have your letter of September 29 asking for further documentation regarding a statement made in my book,** *"The Truth about Water Fluoridation,"* **to the effect that the idea of water fluoridation was brought to England from Russia by the Russian Communist Kreminoff. In the 1930s, Hitler and the German Nazis envisioned a world to be dominated and controlled by a Nazi philosophy of pan-Germanism. The German chemists worked out a very ingenious and far-reaching plan of mass control which was submitted to and adopted by the German General Staff. This plan was to control the population in any given area through mass medication of drinking water supplies. By this method, they could control the population in whole areas, reduce population by water medication that would produce sterility in women, and so on. In this scheme of masscontrol, sodium fluoride occupied a prominent place.**

"Repeated doses of infinitesimal amounts of fluoride will, in time, reduce an individual's power to resist domination by slowly poisoning and narcotizing a certain area of the brain, thus making him submissive to the will of those who wish to govern him.

"The real reason behind water fluoridation is not to benefit children's teeth. If this were the real reason, there are many ways in which it could be done that are much easier, cheaper, and far more effective. The real purpose behind water

fluoridation is to reduce the resistance of the masses to domination and control and loss of liberty."

"When the Nazis under Hitler decided to go to Poland, both the German General Staff and the Russian General Staff exchanged scientific and military ideas, plans, and personnel, and the scheme of mass control through water medication was seized upon by the Russian Communists because it fitted ideally into their plans to communize the world.

"I was told of this entire scheme by a German chemist who was an official of the great I.G. Farben chemical industries, and was also prominent in the Nazi movement at the time. I say this with all the earnestness and sincerity of a scientist who has spent nearly 20 years' research into the chemistry, biochemistry, physiology and pathology of fluorine—any person who drinks artificially fluoridated water for a period of one year or more will never again be the same person mentally or physically." Signed: CHARLES E. PERKINS, Chemist, 2 October, 1954.

Quoting yet another letter by Dr. E.H. Bronner (the nephew of Albert Einstein). Dr. Bronner wrote the following:

Catholic Mirror, Springfield, MA, January 1952:

"It appears that the citizens of Massachusetts are among the 'next' on the agenda of the water poisoners.

"There is a sinister network of subversive agents, Godless intellectual parasites, working in our country today whose ramifications grow more extensive, more successful and more alarming each new year and whose true objective is to demoralize, paralyze and destroy our great Republic—from within if they can, according to their plan—for their own possession.

"The tragic success they have already attained in their long siege to destroy the moral fiber of American life is now one of their most potent footholds towards their own ultimate victory over us.

"Fluoridation of our community water systems can well become their most subtle weapon for our sure physical and mental deterioration. As a research chemist of established standing, I built within the past 22 years 3 American chemical plants and licensed 6 of my 53 patents. Based on my years of practical experience in the health food and chemical field, let me warn: fluoridation of drinking water is criminal insanity, sure national suicide. DON'T DO IT!!

"Even in very small quantities, sodium fluoride is a deadly poison to which no effective antidote has been found. Every exterminator knows that it is the most effective rat killer. Sodium fluoride is entirely different from organic calcium-fluoro-phosphate needed by our bodies and provided by nature, in God's great providence and love, to build and strengthen our bones and our teeth. This organic calcium-fluoro-phosphate, derived from proper foods, is an edible organic salt, insoluble in water and assimilable by the human body; whereas the non-organic sodium fluoride used in fluoridating water is instant poison to the body and fully water soluble. The body refuses to assimilate it.

"Careful, bona fide laboratory experimentation by conscientious, patriotic research chemists, and actual medical experience, have both revealed that instead of preserving or promoting 'dental health,' fluoridated drinking water destroys teeth before adulthood and after, by the destructive mottling and other pathological conditions it actually causes in them, and also creates many other very grave pathological conditions in the internal organisms of bodies consuming it. How then can it be called a 'health plan'? What's behind it?

"That any so-called 'doctors' would persuade a civilized nation to add voluntarily a deadly poison to its drinking water systems is unbelievable. It is the height of criminal insanity!

71

"No wonder Hitler and Stalin fully believed and agreed from 1939 to 1941 that, quoting from both Lenin's 'Last Will' and Hitler's Mein Kampf: 'America we shall demoralize, divide, and destroy from within.'

"Are our Civil Defense organizations and agencies awake to the perils of water poisoning by fluoridation? Its use has been recorded in other countries. Sodium fluoride water solutions are the cheapest and most effective rat killers known to chemists: colorless, odorless, tasteless; no antidote, no remedy, no hope: Instant and complete extermination of rats.

"Fluoridation of water systems can be slow national suicide, or quick national liquidation. It is criminal insanity treason!!" Signed: Dr. E.H. Bronner, Research Chemist, Los Angeles

My question is this: if two extremely intelligent minds from the 1950s state that fluoride in the water is not only damaging to the wellness of every individual person that ingest the deadly toxin (Dr. Bronner called it "treason") why in hell are we drinking and consuming this, and willingly giving it to our kids?

Section 9
Insulin

MSG is known as monosodium glutamate and sodium glutamate. MSG is a sodium salt of glutamic acid. MSG confuses the savory taste part of the tongue with free L-glutamate naturally found in foods. Industrial food manufacturers use MSG as a flavor enhancer because it blends well with other tastes and gives the total perception that it tastes good. Other names for MSG are Ac'cent, Aji-No-Moto, and Ve-Tsin. MSG forces the pancreas to make more insulin, and insulin is a vital hormone in the body. As the body is a fragile balance of hormones, like any delicate ecosystem, not one thing can be out of balance. MSG is linked to making too much insulin, there are many medical conditions are associated with insulin resistance and Type 2 Diabetes.

Type 2 Diabetes:

The pancreas creates too much insulin this leads to insulin resistance, over time, is the body continues to make too much insulin, this can weaken the pancreas. Some individual can no longer produce enough insulin because of a weak pancreas, type 2 diabetes can develop.

Fatty liver:

Fatty liver is strongly linked with insulin resistance. Accumulation of fat in the liver is an appearance of the chaotic mess of lipids that occurs with insulin resistance. Newer research suggests that fatty liver may lead to scarring of the liver and/or liver cancer.

Arteriosclerosis:

Arteriosclerosis, also known as atherosclerosis, is the thickening and hardening of the walls of medium-sized and large arteries. Arteriosclerosis can lead to:

• Coronary artery disease (angina and heart attack)

- Strokes

- Peripheral vascular disease

Other risk factors for arteriosclerosis include:

- High levels of "bad" (LDL) cholesterol

- High blood pressure

- Diabetes

Acanthosis nigricans:

A skin condition linked with insulin resistance that can present as a darkening of the skin in areas where there are creases, such as the neck and armpits.

Skin tags: Skin tags are also seen with increased frequency in patients with insulin resistance.

Reproductive abnormalities in women

- Reproductive abnormalities include difficulty with ovulation and conception (infertility), irregular menstruation, or stoppage of menstruation.

- In contrast, there are no known reproductive abnormalities in adult males with insulin resistance.

Polycystic ovary syndrome:

Polycystic ovary syndrome is a hormonal problem that affects young women. It is connected with irregular periods or no periods at all, obesity, and increased development of body hair.

Hyperandrogenism:

High male hormone levels, which are created by the ovaries, can be seen in insulin resistance and may play a role in Polycystic ovary disease (PCOS), the reason for this is unclear. However, it is thought that insulin resistance somehow causes abnormal ovarian hormone production.

Growth abnormalities:

Some people that have insulin resistance may also experience growth effects from having too much insulin coursing throughout their body. A large amount of people who have high levels of insulin experience weight gain and struggle to lose weight. Other common growth complications from too much insulin can result in a decrease in growth and mental development in children, low bone density and lack of muscle strength.

Have you noticed insulin-related conditions are becoming more common?

Diabetes has been affecting people and animals for thousands of years. An illness that some people think was diabetes was found among the Egyptians in a manuscript dating back to 1500 BC. The National Medical Journal of India, which studies ancient Indians (circa 600 BC), were well aware of the condition. They tested for diabetes, which they called "sweet urine disease" by seeing if ants were attracted to a person's urine. The American Diabetes Association (ADA) reports that in 1910, medical professionals took the beginning steps toward determining a cause and treatment for diabetes. Edward Albert Sharpey-Shafer announced that the pancreas of a diabetes patient was unable to bring about what he termed "insulin," a chemical the body uses to break down sugar. Thus, excess sugar ended up in the urine.

Despite attempts to manage the disorder through diet and exercise, people with diabetes inevitably died prematurely. In 1921, scientists experimenting with dogs had a breakthrough in reversing the effects of diabetes. Two Canadian researchers, Frederick Grant Banting and Charles Herbert Best, successfully extracted the insulin from healthy dogs. They then injected it into diabetic dogs to improve their status.

As recently as twenty years ago, type 2 diabetes was not thought to occur in children. In fact, it was at one time referred to as "adult-onset diabetes," and type 1 diabetes was called "juvenile diabetes." More cases began appearing in children and adolescents over the last two decades due to poor eating habits, lack of exercise, and excess weight.

Despite the steps we have taken since diabetes was first discovered in ancient times, it still remains a major cause of death and health complications. As of 2011, diabetes was the seventh-leading cause of death in the United States, according to the National Diabetes Information Clearinghouse.

Section 10
What is High Fructose Corn Syrup?

High fructose corn syrup is a cheap alternative to normal cane sugar and acts as a food preservative, which is one of the reasons food manufacturers use it. It's used in many foods such as yogurt, cereal, bread, drinks, and even condiments. The Corn Refiners Association has in the past claimed that high fructose corn syrup is natural. Research has found links between high fructose corn and obesity, the rates of child obesity has quadrupled over the past forty years. Child obesity, then leads to obese adults with an astounding risk of developing diabetes, high blood pressure, nerve disease, and cancer.

In 2012 the American Heart Association foresees that over 40.5% of Americans will suffer from some form of heart disease by 2030, the estimated cost of the healthcare is $1 trillion annually. Research suggests that high fructose corn syrup is linked to increasing autism rates. **"To better address the explosion of autism, it's critical we consider how unhealthy diets interfere with the body's ability to eliminate toxic chemicals, and ultimately our risk of developing long-term health problems like autism,"** said Dr. David Wallinga, co-author of the study and physician at the Institute for Agriculture and Trade Policy (IATP).

Many foods that contain GMOs and high fructose corn syrup have pesticides. high fructose corn syrup is an extremely processed sweetener, and it depletes the body of zinc, which is vital for growth and cell division, fertility, and the immune system. Zinc is also used by the body to clean heavy metals like mercury, arsenic, and cadmium, as well as aluminum and other toxins that are known to damage the brain.

People who consume high fructose corn syrup are at risk of developing type 2 diabetes and insulin resistance. In my opinion, high fructose corn syrup is one of the causes of polycystic ovary syndrome (PCOS). PCOS is a disorder that affects an estimated

5-10% of women and is a leading cause of infertility, PCOS has been found in girls as young as 11 years old.

The desensitized cell membrane causes the rejected insulin to stay in the bloodstream, where it causes a variety of conditions, including the unbalanced hormones seen in PCOS and initiates the beginning processes of cardiovascular disease. Unable to get into the cell because of the cell membranes' insulin insensitivity, glucose also stays in the bloodstream and is converted to fat, contributing to weight gain. Eventually, this can lead to numerous serious diseases, such as heart attack or stroke, called metabolic syndrome (syndrome X) and pre-diabetes, which, if ignored, can lead to type 2 diabetes.

If you have PCOS, you should stay away from high fructose corn syrup at any costs. GMO foods can not only prevent you from losing weight, but also counteract the effect of many medication and increase the insulin within your body.

In an article from *The Washington Post* on the 26th of January, 2009: **"Mercury is toxic in all its forms. Given how much high fructose corn syrup is consumed by children, it could be a significant additional source of mercury never before considered. We are calling for immediate changes by industry and the [U.S. Food and Drug Administration] to help stop this avoidable mercury contamination of the food supply,"** the Institute for Agriculture and Trade Policy's Dr. David Wallinga, a co-author of both studies, said in a prepared statement.

26 January 2009 a study published in the current issue of Environmental Health, researchers found detectable levels of mercury in nine of twenty samples of commercial HFCS. And in the second study, the Institute for Agriculture and Trade Policy (IATP), a non-profit watchdog group, found that nearly one in three of 55 brand-name foods contained mercury. The chemical was found most commonly in HFCS-containing dairy products, dressings, and condiments.

In the early 1900s, there was enough cane sugar for every person in the USA to consume 76 pounds every year. About one hundred years ago, people ate very small amounts of corn syrup per capita, and they did not consume high fructose corn syrup. Before the late 1960s, corn syrup was either glucose or dextrose. In the early 1970s, one pound per person. In the early 1990s, there were about 63 pounds per person. It decreased to 50 pounds per person in 2012. The data shown in the chart below starts in the very early 1900s and moving through 2012, stating the per capita of both sugar and high fructose corn syrup. Some research shows that your average grocery store products contain HFCS, and many of the calories from soda are from HFCS.

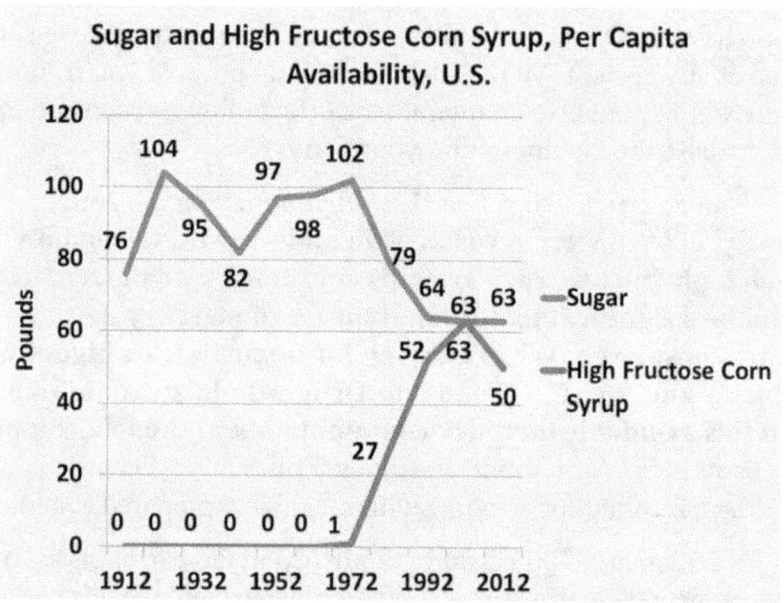

Companies are realizing that the average consumers are becoming aware of the dangers of high fructose corn syrup, so they're changing its name.

Other Names:

Fructose corn syrup; fructose-glucose syrup; glucose fructose syrup; crystalline fructose; fructose glucose syrup; fructose corn syrup solids; maize syrup; glucose syrup;

glucose/fructose syrup; tapioca syrup; dahlia syrup; fruit fructose; crystalline fructose; glucose-fructose; and isoglucose.

Published in Natural Society on May 19, 2014, Christina Sarich states: **"Russian lawmakers are considering equating GMO-related activities to a terrorist act or death threat, a criminal act that Russia says deserves the punishment which killers and creators of mass ecological and human genocide deserve.**

All the research clearly demonstrates that GMO is not only doing harm to mankind, but also causing long-term effects on the environment. Other countries around the world are fighting against GMOs while America and their political leaders are trying to force GMOs on the world's population. On July 23, 2015, the U.S. House of Representatives is trying to pass a bill, H.R. 1599, this bill would prevent any state under federal law from labeling of GMOs on food packaging. H.R. 1599 was created in part by Monsanto's and the "Big Food" industry, the bill was deceptively called "Safe and Accurate Food Labeling Act." also known as the "Dark Act."

Unless you perform a regular detox to remove the GMOs and the GMO poisons, they could be building up deep down in your body and inducing changes to your genetic makeup. This could then contribute to long-lasting illness or even death.

It is very important to frequently cleanse and detoxify your body of gene-altering chemicals that are damaging the natural balance in your intestinal tract, as well as poisoning your blood, causing gastrointestinal upset, and triggering neurological damage, among other conditions.

Here are ways that you can take back your health:

1) **Psyllium seed husks**, also known as ispaghula, isabgol, or psyllium, are portions of the seeds of the plant plantago ovata, (genus plantago), a native of India and Pakistan. They are hygroscopic (retaining moisture), which allows them to expand and become mucilaginous. Since GMOs

have been demonstrated to directly change the bacterial balance within the stomach, psyllium husk is a powerful remedy to help rid the stomach of these transgenic intruders, allowing beneficial bacteria to recover their true place as regulators of the digestive system.

People with irritable bowel syndrome (IBS), constipation, diarrhea, ulcerative colitis, dyspepsia, and other persistent digestive disorders typically experience dramatic relief by supplementing with psyllium husk. When taken along with ample quantities of water, psyllium husk expands into a jelly-like mass that basically scrubs the intestines clean of toxic buildup.

The FDA has shown a real benefit of psyllium seed husk intake and a decreased risk of coronary heart disease (CHD).

2) **Organic sulfur/MSM**. When it comes to insuring that the liver is functioning at its full detoxification capacity, perhaps there is no nutrient more powerful than organic sulfur, which is known as methylsulfonylmethane (MSM).

A vital factor in detoxification, energy output, cell oxygenation, and immune capacity, organic sulfur has gained "near miracle" status among many health professionals who now acknowledge how a deficiency of this vital nutrient can encourage the toxic buildup inside the body.

1) Improving enzyme and glandular hormone production

2) Increasing flexibility in muscle tissue, reduces muscular pain from waste removal, and cellular regeneration.

3) Helps regulate insulin production

4) Increase production of skin cells and reduces wrinkles

5) Rebuilds colon tissues and helps with parasite removal

6) Helps reverse Parkinson's, osteoporosis, and Alzheimer's

7) Promotes growth of hair and nails

Organic sulfur used to be present throughout the food supply before the days of GMOs, petrochemical fertilizers, pesticides and weed killers, and other "modern" agricultural interventions, but today, it is severely lacking. Supplementing with organic, lignan-based sulfur crystals will help fix your malfunctioning cells, restore healthy oxygen transport, and ultimately facilitate a systemic detoxification process that will advance the elimination of GMO remnants from your body.

2) **Organic probiotics**. GMOs change the bacterial balance within the stomach. Foods rich in probiotics like raw sauerkraut, kefir, kombucha tea, fermented vegetables, yogurt, and blue-green micro-algae are vital for protecting the body from the harmful effects of GMO.

4) **Cascara sagrada (sacred bark)**. Colon cleaning is a must for the body's health, as GMOs will stay in the stomach for a long time. Cascara sagrada is a really powerful and well-known colon cleansing herb. It's rich in a compound known as anthraquinone that instigates the contraction of the intestinal walls, which in turn promotes healthy bowel activity. This, united with its laxative effect, makes cascara sagrada one of the most effective colon cleansing herbs known to man, and one that will help protect your system against GMO damage.

5) **Wild burdock root**. This is a potent blood- cleaning agent that is relatively easy to obtain and simple to learn how to use. An aggressive diuretic, wild burdock root is firm enough to rid the body of even difficult-to-reach toxins, including GMO residues and associated pesticide and herbicide chemicals. Burdock root also helps cleanse parasites, heavy metals, bacteria, and other toxins from the

blood, and is frequently employed to treat chronic bacterial and viral infections such as Lyme disease. Particularly when used in junction with other powerful cleansing herbs, such as dandelion root, red clover flower, aloe vera, cayenne, and garlic clove, burdock root is powerful enough to cleanse your blood supply, liver, skin, and your digestive tract of many harmful toxins, including GMOs.

Section 11
Always Read the Label

It's true that not all organic food is 100% safe. Reading the label of the food you buy and understanding what the chemicals are doing to your body is the best way to ensure healthy living for you and your household.

When are organic products not really organic?

When you buy organic food from a store, but it was made in China, that raises a concern. How can a truly organic product survive the trip around the world if it is truly organic? A news investigation has uncovered that many of the "organic foods" we are purchasing at stores with the label "organic" are actually not organic. The products that are made in China have very small writing on the back of the packaging that most people would easily miss, and have not idea that they are not really buying organic food.

The same can be said for foods that are made or grown in India. One evaluation states that 75% of India's drinking water is contaminated by human and agricultural waste. According to a U.N. study on sanitation, more people in India have access to a cell phone than to a toilet.

Studies have shown that genetically modified crops **"can cause liver and kidney damage."** Considering that 90% of soy and corn grown in the United States is genetically changed, knowing how to read the label and understand the research is the only way anyone and accurately understand the truth about the organic label.

The next part of the book will help shed some light on what is really in the food that we buy. This will act as a starting point for you, the readers, to familiarize yourself the names of the chemicals. There aren't just damaging ingredients in foods that you buy in the stores—the fast-food chains are also guilty of using ingredients that are not only dangerous, but strange. In this section, we'll be looking at some of the top ingredients that are used in everyday foods.

Human hair

Human hair has been found in bread. No, I don't mean a full lock of hair; I'm talking about the chemical makeup of hair. There are many kinds of amino acids that your body needs for healthy day-to-day function, but not all amino acids are what your body needs. L-Cysteine is an amino acid used to extend shelf life in many products such as bread. This form of amino acid can also be found in duck and chicken feathers and cow horns, but most that's used in food comes from human hair. In fact, most of the human hair that's used to make L-Cysteine comes from China. It's gathered from barbershops and hair salons and then sold to make L-Cysteine.If you're looking to avoid L-Cysteine, you can start by buying fresh bread from a local baker, or making it at home. The research I have done doesn't show the presence of L-Cysteine in organic bread, as it is not an additive in flour. There are many fast-food chains that use L-Cysteine. You should avoid McDonald's, Dunkin' Donuts, Burger King, and many others, who all use L-Cysteine as an additive.

Beaver Anal Glands

I'm a lover of everything ice cream, but after I found out about this, I had to rethink my ice cream habits, Vanilla, strawberry, and raspberry ice cream contain beaver anal glands. That's right—the anal glands of beavers. Some of them have shown to contain small amounts of beaver urine secretions. This is called castoreum, which comes from the castor sacs of male and female beavers. The FDA has approved castoreum as a food additive called "natural flavoring." You might not even know that you're eating it at all.

Coal Tar

There is a long list of processed foods that contain dyes. A large number of these dyes are made from coal tar. Yellow #5, also known as tartrazine, is linked to childhood hyperactivity. Any product in the EU that contains tartrazine must have a warning label, the United States has no such label or regulation on tartrazine. In the past, this had brought up very big concern from parents, bloggers started to petition Kraft to remove the dyes from their macaroni cheese product, Kraft removed the dye from some of their products in 2014 that targeted kids ie. character-shaped products.

Silicone Breast Implant Filler

Everyone knows that McDonald's isn't known for its healthy foods. Take, for example, the Chicken McNuggets. The nuggets are only about 50% actual chicken. The other 50% is composed of synthetic ingredients, including dimethylpolysiloxane, a substance used in silicone that can be found in Silly Putty as well as breast implant filler.

Personal Lubricant

I bet you didn't know that when you're eating your favorite salad, it might contain propylene glycerol, which is used in not only in some bagged salads, but also some fast-food salads to keep the lettuce green and fresh. It's also a key ingredient in Astroglide and other personal lubricants.

Dog Vaginal Pheromones

There are several wines, sodas, and juices which contain a preservative and anti-fungal agent called methyl paraben. This is extracted from a dog's vagina when she's in heat.

Section 12: Part: 1
Baby Food

Baby foods contain levels of arsenic, the safety limit for arsenic, according to the World Health Organization, is 0.002 milligrams for every 2.2 pounds of body weight. But in 2011, additional studies showed that even at those low levels of arsenic can cause cancer. As the United States consumes more processed food, there is an upwelling of toxic and cancer-causing contaminants, such as heavy metals like cadmium, in the food that we are giving to our children. Baby foods that contain arsenic, is a very real threat that too much exposure can lead to arsenic poisoning. Many of these toxins exceed the safe amounts set by the US government.

"Baby foods used to wean infants off milk have been found to contain 'alarming' levels of toxic contaminants including arsenic, lead, and cadmium. The findings come as officials at the Food Standards Agency and the European Commission are conducting an urgent review to establish new limits for the long-term exposure of these contaminants in food." -Richard Gray, Science Correspondent, and Alastair Jamieson

9:00 PM BST 09 Apr 2011

"These elements have to be kept at an absolute minimum in food products for infant consumption. In infant foods, the highest concentrations of arsenic in the rice-based foods are of particular concern. Experts now believe there are no safe limits for arsenic, and manufacturers should be making more efforts to remove it from their food."

Would you really exposure your young child to any substance that's extremely toxic and could lead to a lifelong medical condition? The supposed "safe" levels of arsenic exposure are extremely damaging to their development in the most critical part of their life. Dr. Oz stated that arsenic can be found not only in baby foods and formula, but can also be found in juices and water. In his article was posted in September of 2011, Dr. Oz reported his findings from his own tests. Some brands of apple juice had high amounts of arsenic.

In addition, arsenic and other chemical and harmful materials are very present in the water supply. Dangerous arsenic levels are higher for a newborn and toddlers, the full impact of arsenic contact with anyone is by no means trivial. Along with arsenic, other heavy metals like lead, mercury, and cadmium, which were also present in the baby food, pose a genuine health hazard.

To lessen the health risks brought about through arsenic and other heavy metals: **"Breastfed babies instead of giving them baby food. Breastfed babies have been shown to be much healthier than babies fed baby food or formula anyway. Avoid processed food. You can expect higher levels of arsenic or any deadly chemical for that matter if you consume processed food instead of raw, organic food. Buy a water filter. This will limit the amount of arsenic consumed through consumption. Buy a shower filter. Skin is the largest organ in the body and one hot shower can yield the same amount of exposure to harmful chemicals as drinking 8 glasses of unfiltered water. Perform a heavy metal cleanse to flush out the heavy metals that are already in your body."**

Section 12: Part: 2
Baby Formula

Even though there are seven brands of formula on in the market, there are only three companies that produce the organic brands of infant formula in the United States. One of the largest producers of infant formula is PBM Nutritionals, owned by Perrigo. The brand name that they produce the formula is PBM manly makes conventional forms of formula. They also make Vermont Organics, Bright Beginnings, Hain Celestial's, Earth's Best, Whole Foods Market's 365 Organic brand, and Walmart's Parent's Choice brand.

Similac Organic is manufactured by Abbott Laboratories, one of the largest makers of pharmaceuticals in the U.S. Baby's Only Organic produces the formula under the trade name Nature's One. Nature's One Markets Baby's Only Organic formula as a "toddler formula" rather than an infant formula. The reason for this is that Nature's One recognized the dire need for breastfeeding until the age of one. Baby's Only Organic is a family-owned

Children younger than ten months old have immature gastrointestinal systems and are unable to properly digest and absorb the nutrients found in cow's milk. Many of the nutrients in baby formula are not natural or easy to absorb. It's hard to mimic the right mixture of vitamins and minerals that the mother's body knows to give her child, which is why breast milk is always best.

Before we have a look at some recipes, there are some things that we need to talk about. We have to look at the fact that the infant's body changes over time, and thus needs a new set of nutrition at each stage. Find out what ingredients are recommended in homemade formulas that are geared to the age of your child and their dietary needs. Goat's milk is an acceptable source of milk. Rice milk is not recommended for infants, as it has high amounts of arsenic within it that's naturally incorporated in the plant. It also doesn't contain enough calories or fat to promote adequate growth or nutrition. Soy milk is not recommended, as soy-based formulas don't contain the proper nutrients to encourage baby's health properly. Soy causes life-threatening hormonal/fertility problems later in life.

There are many reasons why commercial formulas are not a good choice for your child, many of the formulas use a soy base and contain deadly pesticides, but many of the formula containers have the dangerous chemical Bisphenol A (BPA). According to a new article just out in The Breast Cancer Fund, there may be evidence that links an increasing rate of instances of breast cancer and Bisphenol A (BPA), especially when women are exposed to BPA at a younger age. The Breast Cancer Fund advises that women who are pregnant, plan on becoming pregnant, or nursing exclude BPA use as much as possible. Young children are particularly vulnerable to the long-term health effects of BPA, since their brains and bodies are still developing. Bisphenol A (BPA) has been shown to decrease sperm count in lab studies, as well as impacting testes development. From an article published in *TIME*, Bisphenol A (BPA) may also be linked to causing diabetes, aggressiveness, heart disease.

Bisphenol A (BPA) is an estrogen mimickers, causing a significant hormone disruption and is a probable contributor to early puberty in girls, and ADHD, urogenital abnormalities, and other long-term health problems in both girls and boys. The European Safety Authority found that canned commercial formula is a substantial source of Bisphenol A (BPA) for infants, exposing the child to 0.013MG of Bisphenol A (BPA) per 2lbs of body weight per day!

Keep in mind that not all organic formulas are Bisphenol A (BPA) free.

(BPA is used in tiny amounts in the glass jar lids, but independent testing showed no BPA contamination in the baby food. I would use with caution and at your own discretion.)

Even though the formula is heavily tested and looked at by the Food and Drug Administration (FDA), they still acknowledge the fact that the formula is inadequate to human breast milk.

Homemade Baby Formula:

This recipe uses for an infant that is newborn to about a year or more of age.

Ingredients:

> 1/4 Cup Goat Milk
> 1/2 teaspoon Probiotic
> 3ml Liquid Vitamins (this can be optional during the first week)
> 20 Warm Water

When first introducing the babe to this formula, it is not uncommon that they will spit up every now and then. Goats milk is the closest thing to human breast milk and it is easy to digest by infants.

Note: This should be refrigerated and is good for about 48 hours.

Coconut Milk Formula Ingredients

> 2 Cans Coconut Milk with cream on top

1 cup coconut water
1 teaspoon coconut oil
5 ml liquid vitamins
1/2 teaspoon probiotic
20 oz Warm Water
Note: Mix well in a blender

For infants I would use 10 oz of water for the first few weeks of starting this recipe.

Mass-producing our baby food is not by any means natural. Commercial baby food has a shelf life of nearly a year, while homemade baby foods last about three weeks in the freezer. This shows that commercial baby foods are loaded with preservatives and additives. How safe are they really for your baby?

Additives: Nearly all commercial baby foods contain additives such as ascorbic acid or citric acid that are not necessary and are not ordinarily part of the baby's diet. While water may be added for texture and consistency, it shouldn't be the first ingredient, especially in foods for older infants.

Sweets: Many desserts,breakfast foods, and snack items are very high in sugar and shouldn't be consumed by young children. Some websites and media would have you believe that this is what young kids need for energy, but that's farthest from the truth. Yes, kids burn more energy than an adult. However, kids nowadays are sitting more now than they were in the past, as they have school and homework. Modern foods are what's causing child and adult obesity. This is true even when "natural" sugar is used.

After all the research I've done on some of the top brands of baby formula, the only one that I would give my child, sparingly, would be Baby's Only Organic Formula. This formula has the most natural ingredients and fewer harmful chemicals than any other I've seen. I still encourage breastfeeding. This offers a list of lifelong benefits for both the mother and child.

Many are aware that breast milk contains antibodies, or immune molecules, that are transferred to the baby, giving them immunities to allergies and illnesses that the mother is immune to. It's not just a matter of vitamins, minerals, proteins, and fats that makes the breast milk healthier. There are many things that separate human milk from formula and other animals' milk. Doctors recommend that babies shouldn't have cows' milk until they are at least one-year-old. (Don't believe me ask your doctor.)

The Commission on Nutrition of the American Academy of Pediatrics discourages the use of cow's milk under one year of age. The intestinal lining is slower to develop in some children. Lactose intolerance is rare in infants, but some toddlers and older children can develop diarrhea, bloating, and abdominal pain because of their failure to digest the lactose sugar in milk. Likewise, the allergenic proteins may leak through the irritated intestinal lining into the bloodstream and cause an allergic reaction, such as a runny nose, wheezing, or a red, raised, sandpaper-like rash, particularly along the buttocks. Some children who are allergic to cow's milk may even experience regular ear infections.

Cow's milk is not meant for infant human consumption, it's harder for an infant to digest over human milk. This might lead an allergy later on in life, as the infant's immune system is not yet ready for the high levels of nutrients that cow's milk contains. It doesn't matter what form the cow's milk comes in, whether it's whole, low fat, skim, powdered, or any other form. It's still not recommended for children under one year. Human breast milk contains over 100 natural ingredients that can't be reproduced in the formula. In fact, breast milk also contains substances that may significantly enhance your baby's stomach and support the healthy acceleration of their entire nervous system.

I know many women are unable to breastfeed, so you may want to consider the option of purchasing human breast milk, which is becoming a hot commodity online. There are some cautions with getting human breast milk, but with research, it can be safe. I would also look at making your own formulas many recipes can be found online and that will fit your child's dietary need and age of development.

Section 13
Why Soy is Bad for You

In the remote past, Asian cultures only ate soy that was fermented, the mainstream would have you believe that ancient Asian cultures have been eating tons of soy for thousands of years. According to Kaayla T. Daniel, PhD, ancient Chinese cultures first started eating soybeans about 2,500 years ago. Not long after that, they figured out how to ferment it. But the question is, why did they ferment it?

The Asian cultures somehow knew that soybeans contain toxins, they then avoided soy until they were able to nullify these toxins, this is achieved by the fermentation process. In Asia, soy is only used in very small amounts as a condiment, with pork, and seafood, only in recent years has soy products be consumed in large amounts in an unfermented.

Soy was originally considered an inedible plant, used to fix nitrogen in the soil, even to this day, there are many farming communities that use soy for the purpose of controlling the nitrogen within the soil. Soy was used in tofu in early Asian cultures to promote sexual abstinence. This was due to the phytoestrogens in soy that can lower testosterone levels. The Japanese typically combines it with fish broth and seaweed. These food items naturally contain iodine, helping offset the thyroid-suppressing effects of soy. Soy milk was never used historically by Asians to feed their children, and soy formula was not used in China until 1928.

There are numerous problems caused by soy formulas fed to infants, including difficulty digesting, lack of sufficient nutrients, and toxins. However, one of the biggest problems is the hormonal disruption caused by the isoflavones in soy. Soy products contain enough isoflavones to cause severe disruption to the hormonal systems of infants during a critical period in their development.

Studies suggest that soy is playing a role in the current epidemic of infertility, menstrual, and other reproductive problems in humans. Why would a plant have such an effect on humans and animals? The reason is very simple—it's a defense mechanism of the plant. The soy plants produce more phytoestrogens in the drier years, the animals and humans who would eat the soy plant would have lower reproductive ability, and so there would be fewer animals and humans the following year. As the early monks used soy in tofu to promote sexual abstinence, the current prison systems are using it for the same purpose. There have been numerous lawsuits filed against the use of soy in the prison system. If you think back to Nazi Germany, the Nazis used fluoride and tested many forms of fertility control in their internment camps.

"People will have [GM] Roundup Ready soy whether they like it or not," Monsanto spokesperson in Britain, Ann Foster, said. **"The politics of food"**, Maria Margaronis, The Nation, 27 December 1999

Are the United States prisons the new internment camps?

Starting around the early 2000s the prisons started replacing the food that was given to inmates with a soy-based diet. The federal government encourages the production of soy, and most of the soy that's grown in the USA is GMO. Soy is not a nutritious food, and consuming it in considerably large quantities as we do in the United States can cause severe health problems. According to Kaayla T. Daniel, PhD, author of the book *The Whole Soy Story: The Dark Side of America's Favorite Health Food*, **"Thousands of studies link soy protein to digestive distress, thyroid damage, reproductive problems, infertility, ADD/ADHD, dementia, and even heart disease and cancer. Populations at special risk are infants on soy formula, vegetarians who consume soy protein as meat and dairy substitutes, and adults self-medicating with soy foods because of their belief that soy can prevent heart disease and other health problems."**

Even though there are numerous studies that have shown the extreme damage of soy, it is used as an essential food for many prison meals. The prison population is a perfect test market for the soy industry. **"In 2009, a group of Illinois prisoners sued the Illinois Department of Corrections in federal court for endangering their health by feeding them soy, most of which it receives from grain giant Archer Daniels Midland. The suit was funded by the Weston A. Price Foundation. See the 2009 Chicago Tribune article, Soy in Illinois Prison Diets Prompts Lawsuit over Health Effects."**

An inmate, Eric D. Harris, is serving a life sentence at the Lake Correctional Institution in Florida for sexual battery on a child. Harris is suing the department of correction for the high soy concentration he has been eating. **"Excessive soy can be toxic to the thyroid gland,"** said Sally Fallon Morell, the president and treasurer of the Weston A. Price Foundation, a nonprofit group that advocates a diet of whole, largely unprocessed foods and food high in saturated fats, and is publicizing the lawsuit. **"It can have hormonal effects."**

Mr. Harris is not the only inmate who's fed up with the soy diet the inmates are eating. There are nine other inmates at the Danville Correctional Center in Illinois who filed a related lawsuit back in 2009. According to the Weston A. Price Foundation, who was funding the lawsuit, **"Beginning in January 2003, inmates began receiving a diet largely based on processed soy protein with very little meat. In most meals, small amounts of meat or meat by-products are mixed with 60-70 percent soy protein; fake soy cheese has replaced real cheese; and soy flour or soy protein is now added to most prison baked goods."** Since the diet began, inmates have complained of a **"deliberate indifference"** to **"serious health problems caused by so much soy."**

Washington, DC, June 26, 2012 — Plaintiffs in the lawsuit Harris et al. v. Brown: **"Plaintiff's health complaints include chronic and painful constipation alternating with debilitating** diarrhea, vomiting after eating, sharp pains in the digestive tract, especially after consuming soy, passing out, heart palpitations, rashes, acne, insomnia, panic attacks, depression and symptoms of hypothyroidism, such as low body temperature (feeling cold all the time), brain fog, fatigue, weight gain, frequent infections and thyroid disease."

In my opinion, soy is linked to many of the health problems that people in the United States have been experiencing over the last several years. As many of the soy-based food are given to our children at very early ages, this is the start of lifelong health complications.

"Effective January 1, 2006, foods covered by the FDA labeling laws that contain soy must be labeled in plain English to declare that it 'contains soy.' However, there are many foods and products that are not covered by FDA allergen labeling laws, so it is still important to know how to read a label for soy ingredients."

Soy lecithin is produced from soybean oil, which is commonly created through a chemical process using hexane. If a product is licensed non-GMO, you can accept that the soybeans used are not being genetically modified.

There are products that are not required to state if they contain soy.

Edamame (soybeans in pods)	Hydrolyzed soy protein
Kinnoko flour	Kyodofu (freeze dried tofu)
Natto	Okara (soy pulp)
Shoyu sauce	Soy albumin
Soy bran	Soy concentrate
Soy fiber	Soy flour
Soy formula	Soy grits
Soy milk	Soy miso
Soy nuts	Soy nut butter
Soy protein, soy protein concentrate, soy protein isolate	Soy sauce
Soy sprouts	Soybeans
Soybean granules	Soybean curd
Soy lecithin*	Soybean paste
Supro	Tamari
Tempeh	Teriyaki sauce
Textured soy flour (TSF)	Textured soy protein (TSP)
Textured vegetable protein (TVP)	Tofu
Yakidofu	Yuba (bean curd)

*Products that are covered by the FDA labeling laws and contain soy lecithin must be labeled "contains soy."

May Contain Soy:

   Artificial flavoring

   Asian foods (e.g. Japanese, Chinese, Thai, etc.)

   Hydrolyzed plant protein

   Hydrolyzed vegetable protein (HVP)

   Natural flavoring

   Vegetable broth

   Vegetable gum

   Vegetable starch

"English labeling rules: foods that are not regulated by the FDA, cosmetics and personal care products, prescription and over the counter medications or supplements, pet food, toys and crafts."

The FDA has exempted soy oil from being labeled as an allergen, soy that has been hydrogenated, which reduces the amount of polyunsaturated fats and adds trans fats to your body. Trans fats are linked to heart disease and plaque buildup in arteries. Soy causes to cancer by interfering with enzymes your body uses to fight cancer, chronic health problems such as obesity, asthma, autoimmune disease, and bone degeneration. It's linked to diabetes by interfering with the insulin receptors in your cell membranes, among women with underlying coronary heart disease, eating trans fats increased the danger of sudden cardiac arrest. It can lead to decreased immune function, by reducing your immune response. Additionally, it can increase blood levels of low density lipoprotein (LDL), also known as "bad" cholesterol, while lowering levels of high density lipoprotein (HDL), or "good" cholesterol.

Soy is linked to causing reproductive problems by interfering with enzymes needed to produce sex hormones, interfering with your body's utilization of good omega-3 fats. In relation to breast, colon, and prostate cancer, it is well recognized that increased levels of estrogen can be a major contributor. This is why the improper and overuse of soy products is such a problem. Soy has been shown to contribute to the physical **"feminizing"** of young men and the early development of young girls and boys.

Section 14
Wheat and Grain

In the past, I have been asked by people who are on a weight-loss diet if wheat is okay to eat, and why is it so bad for you. I have decided to do some painstaking research and help debunk the wheat dilemma.

In the case of wheat, I don't know if organic would be any better than GMO. The facts I've found are not only exciting, but shocking in many ways. From the research I have done, there are six classes of wheat, including the spring and winter types. The classes are then determined by the hardness, color, and the time it was planned.

Per 100 g	Einkorn Wheat	Hard White Wheat	Soft White Wheat	Hard Red Winter Wheat	Hard Red Sprint Wheat	Soft Red Winter Wheat	Wheat Durum
Proximates							
Protein (g)	18.20	11.31	10.69	12.61	15.40	10.35	13.68
Vitamins							
Thiamin (mg)	0.50	0.39	0.41	0.38	0.50	0.39	0.42
Riboflavin (mg)	0.45	0.11	0.11	0.12	0.11	0.10	0.12
Niacin (mg)	3.10	4.38	4.77	5.46	5.71	4.80	6.74
Vitamin B-6 (mg)	0.49	0.37	0.38	0.30	0.34	0.27	0.42
Carotene, beta (µg)	19.00	5.00	5.00	5.00	5.00	0.00	no data
Carotene, alpha (µg)	53.00	0.00	0.00	0.00	0.00	0.00	no data
Vitamin A (IU)	312.00	9.00	9.00	9.00	9.00	0.00	0.00
Lutein +zeaxanthin (µg)	769.00	220.00	220.00	220.00	220.00	0.00	no data
Minerals							
Iron (mg)	4.59	4.56	5.40	3.19	3.60	3.21	3.52
Phosphorus(mg)	415.00	355.00	402.00	288.00	332.00	493.00	508.00
Potassium (mg)	390.00	432.00	435.00	363.00	340.00	397.00	431.00
Zinc (mg)	2.24	3.33	3.50	2.65	2.78	2.63	4.16
Manganese (mg)	4.40	3.82	3.40	3.99	4.06	4.39	3.30

Source: einkorn.com

So why do people say wheat is bad for your health?

Humans all over the world have been consuming wheat for thousands of years. Is it really the wheat or is it what modern man is doing to the wheat crops? Farmers fight back against Monsanto has just been instigated by Ernest Barnes, a wheat farmer in Morton County, Kansas. He filed suit back in 2013 in the U.S. District Court in Wichita, Kansas, "alleging that Monsanto's genetic pollution is financially damaging himself and other farmers."

Barnes' court case seemed to be well supported by the facts: that in late May 2013, the USDA announced the shocking discovery that GMO wheat from Monsanto's "open field" experiments had supposedly escaped and spread into commercial wheat farms. Just after that, Japan and South Korea cancelled wheat purchase contracts from the United States, and more cancellations are expected to follow.

For the record, Mike Adams, NaturalNews, openly warned about this possibility in a 2012 article called, *"Stop Out-of-Control Science."* There, Mike Adams wrote: **"Humanity has reached a tipping point of developing technology so profound that it can destroy the human race; yet this rise of "science" has in no way been matched by a rise in consciousness or ethics. Today, science operates with total disregard for the future of life on Earth, and it scoffs at the idea of balancing scientific "progress" with caution, ethics or reasonable safeguards. Unbridled experiments like GMOs have unleashed a self-replicating genetic pollution that now threatens the integrity of food crops around the world, potentially threatening the global food supply."**

Many Americans don't realize that GMO foods are banned from nearly every country around the world. USA still uses GMO crops and has managed to avoid being labeled. Many of the modern wheats contain numerous proteins are left indigestible. These proteins have the potential to cause many health problems in people who consume them. Mike states that: **"Traditional heritage wheat varieties, on the other hand, such as Kamut (khorasan), spelt (dinkel), and Einkorn are said to be more tolerable among individuals with gluten sensitivity, as they contain fewer and different types of gluten. It is still important to practice caution, though, even when trying these ancient wheat varieties."**

"What we found is that the molecules created in this wheat, intended to silence wheat genes, can match human genes, and through ingestion, these molecules can enter human beings and potentially silence our genes," said Heinemann in a press conference on the threat of GM wheat. "We found over 770 pages of potential matches between these two genes in wheat and the human genome," he continued. "We found over a dozen matches that are extensive and identical, and sufficient to cause silencing in experimental systems. The findings are absolutely assured. There's no doubt that these matches exist. From this information, we know that it's plausible, there will be an adverse effect and therefore that's why we're calling for a particular battery of experiments to be done before humans eat this wheat."

Judy Carman continues, "Before this comes near any human feeding studies, you need to undertake thorough animal safety assessments, where you actually look to see if the animals get sick. So you need to see if this genetic modification survives digestion and gets into the bodies of the animals. You need to see what effect it has on them. You need to do proper long-term toxicology studies. You need to check for cancer, you need to see if there are any reproductive problems, and you need to check for allergies…"

So is it the wheat, or is it the genetically modified gluten that is bad for you?

Studies conducted on autism, researchers believe that a link for other behavior-altering disorders have been found. In a new study published in the open-access journal PLOS One, the researchers from Columbia University in New York found that wheat, more predominantly wheat gluten, can be the catalyst that triggers a unique immune response in autistic children. This response is stronger in people who have gastrointestinal problems. The result from the effect mimics the symptoms commonly associated with autism. "It has been theorized that when the immune system forms antibodies against gliadin, these antibodies cross react with self-structures within the nervous system," writes Sayer Ji for GreenMedInfo.com about the process.

"Known as molecular mimicry, this breakdown of immunological self-tolerance can contribute to a wide range of neurological problems, including neuropathy, ataxia, seizures, and neurobehavioral changes including mania, schizophrenia and autism…Anti-gliadin antibodies are therefore a possible cause of autoimmune neurological damage."

Some people suggest that many of the symptoms demonstrated by individuals who are diagnosed with autism could be a direct result of an antibody response caused by overexposure to the gluten within wheat. Nevertheless, wheat gluten seems to increase the levels of the anti-gliadin antibody within the body.

"A subset of children with autism displays increased immune reactivity to gluten, the mechanism of which appears to be distinct from that in Celiac disease," wrote the authors in their conclusion. **"The increased anti-gliadin antibody response and its association with GI symptoms points to a potential mechanism involving immunologic and/or intestinal permeability abnormalities in affected children."**

There is a large amount of toxicity that's contained within wheat gluten. This might be one of the causes of Celiac disease and other autoimmune diseases. Reactions are mostly from peptides contained in gluten known as gliding proteins, which harm the intestinal tract. If Celiac disease left undetected leads to malnourishment, that can lead to other diseases. Gluten toxicity can result in many other malnourishment-related health problems, including chronic fatigue and mental disorders. Wheat is not the same today as it was fifty years ago. It has been crossbred and changed over time to resist fungus, grow more quickly, and be more flexible for industrial bread baking. Only fifty years or so ago, wheat did not contain the amount of gluten that it does today, which is about 50%. It only had 5% gluten. Agriculturists had hurried up the hybrid procedure to accommodate the baking industry mechanical requirements of flexible proteins, leading to the tenfold increase of wheat gluten since that time.

In this section, we have seen that it is not only the wheat, but the gluten from the GMO wheat crops, that is making us sick. Is organic a better option, or are wheat and wheat gluten best to avoid altogether?

The

Gluten Free
Amaranth: Add amaranth to gluten-free flour, sauces soups and stews. this gluten-free is a thickening agent. Amaranth is very high in iron, calcium, lysine, protein, and magnesium with its low carbs content it makes one of the best gluten free grains.
Buckwheat: This grain is not part of the wheat family it is commonly used in waffle mixes and muffins, buckwheat has many nourishes minerals and dietary soluble fiber and is loaded with antioxidants.
*Oats: This grain is naturally gluten-free (only in its purest form), if you suffer from Celiacs only purchase "Gluten-Free" oats. Oats are great in regulating blood sugar and high in dietary fiber.
Corn/polenta: Is a low carbohydrate food that is rich in vitamin A and C, it is also a great source of lutein and zeaxanthin.
Quinoa: This great rice substitute has essential fatty-acids and is high in protein and minerals.
Rice: Wild rice is a great source of dietary fiber and vitamin B, manganese, selenium, and iron. Both wild and brown rice in place of white rice that has been refined and processed.
Teef: This grain has in used for hundreds of years, it is packed with nutrition such as protein, calcium, and iron.

debate regarding wheat isn't looking good so far, but the proof is in the gluten. Traditional wheat only contains 5% gluten. Purchasing organic heritage wheat will cut the bad gluten by up to 45%. Modern wheat has been linked to reducing the blood flow to the frontal cortex of the brain.

Some research is stating that modern wheat might be the cause of gluten intolerance. In a film by Jeffrey Smith and Tom Malterre, MS, CN Discuss GMOs and Gluten from "Whole Life Nutrition," the film talks about the effects of GMO gluten. It can cause a wide variety of health problems.

Digestive disorders have skyrocketed since GMOs have entered the food supply. Irritable bowel syndrome, ulcerative colitis, chronic constipation, gastrointestinal infections, Crohn's disease, leaky gut syndrome, and acid reflux have all seen sharp inclines in the last two decades, as has the incidence of gluten intolerance and Celiac's disease.

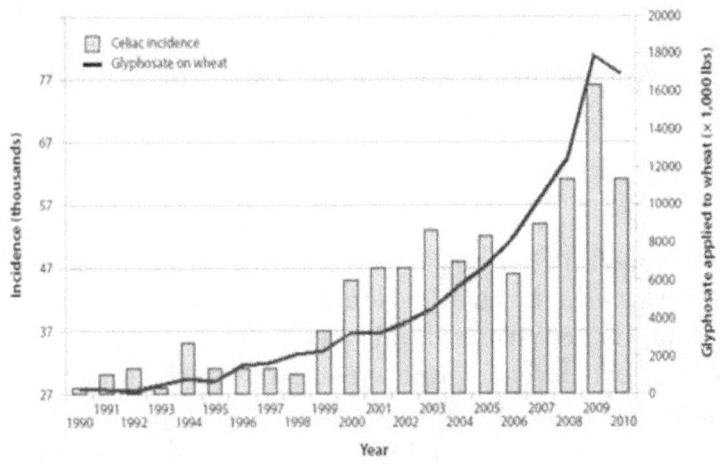

Figure 1. Hospital discharge diagnosis (any) of celiac disease ICD-9 579 and glyphosate applications to wheat (R=0.9759, p<1.862e-06). Sources: USDA:NASS; CDC. (Figure courtesy of Nancy Swanson).

Monsanto was one of the first to successfully create genetically modified plant cells. They started creating GMOs back in 1983. Then they conducted field trials of GMO crops in 1987. Even now, if you were to search for "GMO wheat" online, you would get a multitude of misinformation, and it's very hard to tell what is real.

An article alert on the Greenpeace website states: **"Monsanto's unapproved genetically engineered (GE) 'Roundup Ready' wheat was found growing in a random**

Oregon field last week. The farmer doesn't know how it got there. Neither does anyone else, since Monsanto ended field testing this type of wheat eight years ago. The US Department of Agriculture (USDA) is currently investigating the extent of the contamination. But this story isn't surprising. GE crops can't be controlled. The fact is that contamination happens all the time while companies like Monsanto experiment with nature and our food supply. Putting an end to field testing is the only way to stop it."

If you were to research wheat online there are articles calming how wheat is "not" GMO, but in fact, it has been hybridized. Either way you say it, whether it's GMO or hybridized, it's still not natural wheat. The answer is clear—modern wheat is not healthy for you. The gluten that's produced for industrial standards is causing health problems.

Is organic any different?

A study conducted by Organic Center has shown that there are very compelling reasons to eat organic wheat. **"Grown without pesticides, it may develop more robust chemical defenses against environmental stresses and predators—and many of the protective compounds acts as antioxidants,"** says Erin Smith, senior science consultant with the Organic Center. **"Deeper root systems also allow the plants to draw more minerals from the soil, and organic farmers tend to plant older or native varieties, which are frequently more nutritious."**

Organic wheat, although still containing wheat gluten, has less of a form of gluten than modern wheat. Early organic forms of wheat, like Einkorn and Kamut, are mostly grown organically, and many people with celiac disease and gluten sensitivity have no problem consuming these forms of wheat. The gluten within these grains is less toxic than the gluten in modern wheat crops.

Section 15
Rice is it Really Healthy?

Is it the super food everyone thinks it is? The real danger is not that rice contains arsenic naturally, but the fact that it is at very high and dangerous levels. New research done by *Consumer Report* has found that many types of rice-based products contain arsenic. This toxic carcinogen is linked to causing cancers such as lung, liver, kidney, prostate, and skin cancer. As many baby foods use rice as a base ingredient, arsenic is leading children to long-term health problems later in life. At the present time, there is no federal limit for arsenic in foods. The level that's accepted in drinking water is 10 parts per billion (ppb). *Consumer Report* has shown that 1 serving of rice could give an adult close to one and a half times the safe amount of inorganic arsenic. This would equal drinking about one liter of water.

The study from Consumer Report states:

"White rice grown in Arkansas, Louisiana, Missouri, and Texas, which accounts for 76 percent of domestic rice, generally had higher levels of total arsenic and inorganic arsenic in our tests than rice samples from elsewhere. Within any single brand of rice that was tested, the average total and inorganic arsenic levels were always higher for brown rice than for white. People who ate rice had arsenic levels that were 44 percent greater than those who had not, according to our analysis of federal health data. And certain ethnic groups were more highly affected, including Mexicans, other Hispanics, and a broad category that includes Asians."

As stated earlier in this book, the word "natural" doesn't mean that it's safe. According to the federal Agency for Toxic Substances and Disease Registry: **"The U.S. is the world's leading user of arsenic, and since 1910, about 1.6 million tons have been used for agricultural and industrial purposes, about half of it only since the mid-1960s. Residues from the decades of use of lead-arsenate insecticides linger in agricultural soil**

today, even though their use was banned in the 1980s. Other arsenical ingredients in animal feed to prevent disease and promote growth are still permitted. Moreover, fertilizer made from poultry waste can contaminate crops with inorganic arsenic." A study done in 2009 by the EPA estimated that rice contributes to only 17% exposure to inorganic arsenic, fruits and fruit juices at 18%, and vegetables at 24%, this means that you can be consuming poison. Although rice only contributes to 17% of inorganic arsenic, cereal products with rice-based ingredients expose people to about 50% of the inorganic arsenic.

Why does rice have arsenic in it, anyway?

Rice requires a very wet environment in order to grow, this in turn allow arsenic to be more easily taken up by the rice roots and stored in the grain. In studies performed by several scientific bodies, inorganic arsenic levels were higher in the brown rice than in the white rice. The process of making white rice is by "polishing" the surface and removing some of the surface layer. Doing that will somewhat reduce the amount of inorganic and organic arsenic contained within the rice grain. Is organic rice any better? Well, the answer is yes and no. Rice has arsenic in it naturally, so organic rice will contain arsenic. However, the organic rice contains naturally occuring arsenic, and not the inorganic arsenic that modern rice would contain. As organically grown rice has to follow strict rules, it shouldn't contain any pesticides, one of the leading causes of high levels of arsenic in the food. Going organic not only for your rice products, but also for the other foods, would help reduce the levels of arsenic in your daily diet. One of the ways to help your body fight off the arsenic in your body is by taking iodine, which could help reduce the heavy metals that can bind to arsenic.

Section 16
The Oil Debate

Many people don't know which oil is healthy for you, and what it's really used for. In the past, oils were not used the way they are today. There are literally hundreds of edible oils—it's no wonder there's confusion. In this section, we'll have a look at the top oils.

The media would have you believe that vegetable oils are very healthy for you, but the research is stating otherwise. Vegetable oils are actually known as seed oils, and these kinds of oils turns into dangerous trans fats. Trans fats have been proven to cause brain atrophy, as well as heart disease, cancer, obesity, and diabetes. Trans fats are also commonly found in processed foods. If the food is labeled "hydrogenated" or "partially hydrogenated," that means trans-fat. Many countries around the world have banned vegetable oils due to the link between its use and health risks.

The risk of trans fats is so real that it was banned in New York back in 2006, with the full ban to be in effect by July 2008. The FDA has stated that they, too, have even thought about banning trans fats in the form of hydrogenated oils. Trans fats have been linked to chronic inflammation, if you have allergies or arthritis, dermatitis, or colitis, you might be consuming too many trans fats. Chronic inflammation is one of the leading causes of heart disease, cancer, chronic lower respiratory disease, stroke, Alzheimer's, diabetes, and nephritis.

In today's normal conditions, we all consume on average 9% of our calories from these types of oils.

Coconut Oil. Even though coconut oil contains fats, as do all other oils, these kind of fat is actually very healthy. Because of the balance of coconut oil, it makes an enormously stable meaning it can take high heat making a great cooking oil. Coconut oil contains 50% lauric acid. This kind of fatty acid is a natural antibacterial, antiviral, and antifungal. The regular use of coconut oil lowers the risk of heart disease by increasing good cholesterol, and is found to enhance the immune system. Lauric acid is so healthy for you that human breast milk contains this amazing natural acid. The only other way of getting lauric acid in your diet, other than consuming breast milk, is by coconut. The other benefit of coconut oil is that it contains good chain fats that supply energy directly to the brain. People with insulin resistance or who are diabetic will have no increase of insulin. Some researchers are even looking at it in the treatment of Alzheimer's and other neurological disorders.

Grass Fed Butter. According to the media, butter is very unhealthy, but that could not be further from the truth. Butter contains all of the fat-soluble vitamins (vitamins D, E, and K) and is a very good source of vitamin A. It's very rich in trace minerals that your body needs, and grass-fed butter even has omega-3 fatty acids. The real key benefit from butter is the high butyrate content. Butyrate is a fatty acid that's linked to many health benefits. Butter is very easy on the digestive tract, and even reduces chronic inflammation. It fights against neurodegenerative disorders like Alzheimer's and Parkinson's. Many people think that butter will make you fat, which is not true. Organic butter is linked to weight loss by means of balancing your insulin and stabilizing blood sugar levels. Organic grass-fed cow's milk that's made into butter has high levels of linoleic acid, a compound that protects against cancer, reduces inflammation, and makes it easier to burn fat and retain muscle mass.

The Truth about Canola Oil, have you ever heard of a canola plant? If you do your research, you'll find a big surprise. Canola actually stands for "Canadian oil" It's created from a GMO form of rapeseed. They changed the name because people started to catch on that it was not healthy. This as well contains trans fats that are dangerous to your health.

Many of the so-called "healthy" vegetable oils have very high amounts of biologically active fats called omega-6 polyunsaturated fatty acids. This doesn't apply to healthy plant oils like olive oil or coconut oil, which are extremely good for you. It's true that your body does need omega-6 fatty acids. However, it's essential that you keep this in a delicate balance. As your body can't produce omega-3 or omega-6 fatty acids, we have to get them from our diets. The ratio of omega-6 and omega-3 in the typical diet today is extremely high—about 16:1, on average.

This, of course, differs with each individual. Omega-6 has increased three times in our diet over the past fifty years. The research suggests from both randomized controlled trials and observational studies that vegetable oils can increase the risk of cardiovascular disease.

High oleic sunflower oil. This oil presents a possible risk for those who suffer from diabetes and insulin resistance.

According to WebMD: **"High oleic sunflower oil mildly increases blood sugar and influences insulin levels. Therefore, people with type 2 diabetes can experience the side effect of developing hardening of the arteries. Several studies prove that allergic reactions to sunflower oil are very rare cases. Their main symptoms can be swelling, skin rashes and itching, stomach cramping, nausea and vomiting. Highly refined oil usually doesn't provoke allergy."**

According to Journal of the National Cancer Institute (2001), **"High levels in the body of both oleic and monounsaturated fatty acids are associated with the increased risk of breast cancer."**

Some research has shown that sunflower oil omega-6 fatty acids accelerate the growth of human prostate tumors. Omega-6 fatty acids are found in vegetable oils including corn, canola, soybean, and sunflower. The researchers pointed out that the rate of prostate cancer in the United States has increased steadily along with intake of omega-6, signifying a possible relation between omega-6 fatty acids and prostate cancer. Excessive consumption of these vegetable oils can lead to:

• Asthma

• Blindness

• Heart disease

• Cancer

Olive Oil: There are dangers when it comes to the use of cooking with olive oil. Many scientific studies have shown repeatedly that olive oil is one of the best, healthiest oils for the many varieties of the omega fatty acids your body needs for a healthy nervous system and cardiovascular system. The problem with using olive oil for cooking is that the process of heating oils can cause the fats to become carcinogenic, which means you're increasing your risk of cancer. Heating up the oil causes enzymes to be destroyed, proteins are denatured, fats become carcinogenic, the sugars become caramelized, and vitamins and minerals become less available.

Much of the olive oil that's on the supermarket shelves is not 100% olive oil. They might be sold as "Italian olive oil," the olives aren't even grown in Italy. They are in truth grown in countries like Spain, Morocco, and Tunisia. The olives are then picked and shipped to Italy, which is the largest importer of olives. At the processing plant, lesser oils are mixed in, such as soy oil.

This doesn't happen in all of the refineries—just some. Some of the refineries are mixing vegetable oils with beta carotene create imitation olive oil. The bottles are then labeled "extra virgin" and claim to be from Italy. 69% of the oils that are shipped are not even able to pass an expert taste and smell test for real olive oil.

One way to test if you're really getting the authentic olive oil is by looking for oils that are sold in dark glass or metal containers, a brand that I like is Agiolia early harvest organic olive oil. Heat, light, and oxygen are the number-one enemies of olive oil, as they increase the decay. Do not get olive oil that is in clear glass or plastic. If you got real olive oil, it would be better tasting and fresh. Store in a cool, dark place, not on the counter or in the cupboard.

Palm kernel oil. This oil is a yellowish color and is derived from the kernels of a palm. It consists of 75% saturated fats and is without a doubt one of the unhealthiest oils used for cooking. Palm olein is the liquid portion of palm oil obtained when the oil is separated by a process called fractionation. While palm oil has no trans fats, it does carry high levels of saturated fats, which can be a substantial threat to cardiovascular health.

Palm oil. When this oil is unchanged, it's trans-fat free. However, it's about 80% saturated fat palm oil and contains about twenty-two grams of saturated fat per two tablespoons. As per the 2,000-calorie-per-day diet, that's the maximum amount of saturated fat you should ingest, and experts agree that saturated fats raise the levels of "bad" cholesterol within the blood. This could lead to damage in the heart and arteries, possibly leading to a heart attack or stroke. The buildup of some kinds of saturated fats causes palmitic acid. This acid is the core fat in palm oil. Research have shown that palmitic acid caused lab mice to resist appetite-suppressing hormones such as leptin and insulin. The oil obtained when the oil is separated by a process called fractionation. The researchers believe that this could make you eat more, as your body doesn't get the signal to stop eating.

Cottonseed oil. This is mostly made with GMO crops. The Institute for Responsible Technology reported that this presents a risk to the workers in the fields.

Workers exposed to Bt cotton developed allergies:

> 1. Agricultural laborers in six villages who picked or loaded Bt cotton reported reactions of the skin, eyes, and upper respiratory tract.

> 2. Some laborers required hospitalization.

> 3. Employees at a cotton gin factory take antihistamines every day.

4. One doctor treated about 250 cotton laborers.

Sheep died after grazing in Bt cotton fields.

1. After the cotton harvest in parts of India, sheep herds grazed continuously on Bt cotton plants.

2. Reports from four villages revealed that about 25% of the sheep died within a week.

3. Postmortem studies suggest a toxic reaction.

Coton oil is made by means of hydrogenation, which increases the shelf life of the oil. Hydrogenated oils are chemically altered by forcing hydrogen into the oil at enormously high temperatures. This converts the fats within into dangerous trans fats, which are linked to diabetes, heart disease, and obesity. As the cotton is made with GMO seeds, it will also contain pesticides.

Peanut oil. In the United States, many of the peanuts that are used are GMO, and the oil that's then produced contains about 48% oleic acid, 18% saturated fat, and 34% omega-6 linoleic acid. If you're going to use peanut oil, only buy and use organic. This will cut out the pesticides, if not stored properly, peanut oils can grow a mold and fungus. This could result in aflatoxin toxin poisoning.

Corn oil. About 85-90% of the corn grown in the United States is currently GMO, containing Bt. The oil is somewhat higher in saturated fats than olive oil, but is rich in omega-6 fatty acids. Corn oil is unhealthy because it is highly hydrogenated to extend its shelf life, and it lowers both the good and bad cholesterol. I wouldn't use or consume corn oil as is it very unhealthy.

Soy oil. We have talked in depth about the dangers of soy, so there is not much more to say about soy oil. It's still soy.

Section 17
Milk!

There are many false claims made about milk, some by the media, and others by the manufactures of the products themselves. This in not only limited to whole milk, but also non-homogenized and unpasteurized. The mainstream dairy farms would have you believe that the word "organic" only means more money—that is not true. Organic milk comes from cows who only eat organic feed. Organic milk doesn't contain a hormone cocktail including pituitary steroid, hypothalamic, and thyroid hormones.

Additionally, organic milk shouldn't contain growth inhibitors MDGI and MAF, and shouldn't have rBGH (recombinant bovine growth hormone) injected into cows to increase milk production, rBGH is genetically engineered hormone directly linked to breast, colon, and prostate cancer. Much of the non-organic milk with growth hormones might be one of the reasons that our children are experiencing puberty earlier.

Many people who suffer from milk-related conditions such as lactose intolerance are, for some reason, able to drink organic milk without any further problems. This is true with many organic foods. Some studies indicate that organic milk has higher level of omega-3 fatty acids and antioxidants. Organic dairy farmers are required by law to follow a strict set of rules, and the cows are required to spend time on organic pastures.

"Organic production is more environmentally sound because organic farmers are not using toxic and persistent synthetic pesticides that can remain in the air, water or soil for years to come," says Holly Givens, public affairs manager of the Organic Trade Association, based in Greenfield, Mass.

Homogenized and non-homogenized milk were once called **"the worst thing dairymen have done to milk."** Homogenized milk is made through the process of mixing substantial quantities of milk, in most cases from several different milking herds or dairies.

Raw milk will normally and naturally separate into layers of fat density. Homogenization is a process that's done mechanically by forcing the milk at high pressure through small holes. During the process of pressing the milk, you change the size of the fat globules. Some speculate that this will also rearrange the fat and protein molecules. In turn, this might affect how the human body reacts. In the 1970s, Kurt Oster proposed the hypothesis that homogenized milk might increase your danger of nerve disease. The evidence that drinking homogenized milk is harmful to humans is inconclusive—no research to date has validated Dr. Oster's hypothesis. Most consider the question of homogenized milk causing heart disease to be no more than myths and rumors.

Grass milk doesn't mean that it's made from grass. This is the name of milk that's from grass-fed only cows. Research shows that this milk is high in omega-3 fats, vitamin E, beta carotene, calcium, and conjugated linoleic acid.

Raw milk is surrounded by a lot of controversy. Raw milk farms have been shut down in many states in the USA. Why? Humans have been drinking raw milk for thousands of years. Research shows that a single glass of milk can contain a cocktail of up to twenty painkillers, antibiotics, and growth hormones. We are giving our children high quantities of medicine in their milk!

WHAT'S IN YOUR PINT?

Chemical	What is it?
■ Niflumic acid	anti-inflammatory painkiller
■ Mefenamic acid	anti-inflammatory
■ Ketoprofen	anti-inflammatory
■ Diclofenac	anti-inflammatory
■ Phenylbutazone	anti-inflammatory
■ Florfenicol	antibiotic
■ Estrone	natural hormone
■ 17ß-estradiol	sex hormone
■ 17a-ethinylestradiol	steroid hormone
■ Naproxen	anti-inflammatory
■ Flunixin	anti-inflammatory
■ Pyrimethamine	anti-malaria drug
■ Diclofenac	anti-inflammatory
■ Triclosan	anti-fungal drug

Source: University of Jaen, Spain

Modern milk also contains the hormone 17-beta-estradiol. This is a sex hormone called estrogen. The research shows that the hormone is found in three millionths of a gram in about 8.6 ounces of milk, while the highest dose of niflumic acid was less than one millionth of a gram per 8.6 ounces of milk.

So how does raw milk compare? Raw milk that comes from organically raised cattle is rich in natural and healthy bacteria, including lactobacillus and acidophilus. There's a lot of fear about getting sick from raw milk. In truth, there are many coliform classes (bacteria) that are found in raw milk. These are good for you. Many people don't know that there are 230 kinds of E. coli, but out of the 230, only two or three are pathogenic and might cause sickness. The rest help boosts your immune system and digestion.

When you pasteurize milk, it destroys the nutrients from raw milk. Without the enzymes that are destroyed during the pasteurization process, the milk becomes very hard for the body to digest. Research has shown that organic raw milk is easily digested by people who are lactose intolerant. Some people even go as far as saying that they have no symptoms at all after consuming raw milk, and their lactose intolerance just disappears.

Also, raw milk contains phosphatase, an enzyme that helps the absorption of calcium in your body and strengthening your bones. All of the enzymes are killed as you cook the milk above 120 degrees, as is done with pasteurization. Raw milk also has something called butterfat, a very good source for vitamin A, and contains acids with strong anti-carcinogenic benefits. Some research suggests that raw milk can even help fight cancer, and has a healthy balance of omega-3 and omega-6 ratios.

NaturalNews on Tuesday, July 02 2013, by Ethan A. Huff:

"While it is clear that there remains some appreciable risk of food-borne illness from raw milk consumption, public health bodies should now update their policies and informational materials to reflect the most high-quality evidence, which characterizes this risk as low," said Ijaz during the presentation. **"Raw milk producers should continue to use rigorous management's practices to minimize any possible remaining risk."**

"The scientific papers cited at the BC Centre for Disease Control presentation demonstrated a low risk of illness from unpasteurized milk consumption for each of the pathogens," explains a press release about the finding. **"This low-risk profile applied to healthy adults as well as members of immunologically-susceptible groups: pregnant women, children, and the elderly."**

In the early 1900s, raw milk was used as medicine. Dr. J.R. Crewe's "Milk Cure" was used at the Mayo Clinic and was effective in the treatment of cancer, weight loss helping your slim your waist, kidney disease, allergies, skin problems, urinary tract problems, prostate problems, chronic fatigue, numerous other

119

chronic conditions **"Hypertension responds with equal gratification. The blood pressure improves rapidly,"** wrote Crew. **"I have never seen such rapid and lasting results by any other method."**

Raw milk is not sold in every state, and there have been reports that the FDA has closed down many raw milk producers nationwide. The reason behind this, I don't know, but it looks as if they don't want the health benefits to get out to the public.

The less work and more natural a product is, the healthier it is. In my opinion, drinking raw milk has been done for thousands, and in many locations around the world, people still drink raw milk. If it was really as bad for you as the FDA claims, humans would have never continued to drink raw milk.

Jersey and Guernsey milk comes from cows bred on the Isle of Guernsey and the Isle of Jersey, which are located off the British mainland. The milk Guernsey cows produce is high in fat, and is also loaded with beta-carotene and protein. Jersey cows produce milk that's high in butterfat, which is why it is often used in cheese-making.

Lactose-free milk is for the many people around the world who suffer from lactose intolerance, but how can milk that contains lactose really be lactose free? Lactose-free milk simply contains lactase, a natural enzyme that helps break down lactose. Some people who take a lot of the lactase enzyme could experience bloating and diarrhea after they have stopped consuming the enzyme. This is from the body trying to break down the lactose without help.

Skim milk is one of the most-consumed milk products in the USA. The media would have you believe that normal milk is far too fattening, with calories and cholesterol that could give you heart disease. Around 1918, consumers were concerned about the fat in their dairy foods. The dairy producers jumped on this new trend. The once "industrial waste" was now the new healthy super milk. Skim milk appears to be an off-blue color, with a chalky taste, and it's very runny, like water. Skim milk producers add powdered milk to thicken it so it looks and tastes more like normal milk. The powdered milk that's added is not required by the FDA to be placed on the label. They justify this by claiming that powdered milk is still milk.

There are some studies that suggest pasteurized milk and most importantly skim and low-fat milk can increase the risk of cancer. Non-organic cow's milk contains rBGH (recombinant bovine growth hormone, also known as recombinant bovine somatotropin), which is a genetically engineered hormone that's injected into cows for the purpose of increasing milk production.

"Bristol, United Kingdom. Several studies have shown powerful associations between blood levels of insulin-like growth Factor-I (IGF-1) and the risk of colon cancer, prostate cancer, and premenopausal breast cancer. As a matter of fact, recent evidence indicates that high IGF-1 levels may be more important than other previously reported risk factors for cancer. IGF-1 is released by human growth hormone and stimulates growth throughout fetal and child development. IGF-1 in the body is normally tightly bound to a large protein molecule (IGF binding protein-3) and there is evidence that high levels of IGF binding protein-3 protect against the development of certain cancers.

The pharmaceutical industry is well aware of the increasingly clear association between IGF-1 and cancer. Chemotherapeutic drugs are being developed to block the activity of IGF-1 or enhance the activity of IGF binding protein-3." A quote from Smith, George Davey, et al. Cancer and insulin-like growth Factor-I. British Medical Journal, Vol. 321, October 7, 2000, pp. 847-48 (editorial)

1% and 2% fat milk have been shown in studies to increase the risk of developing prostate cancer. The first study consisted of 82,000 men aging from 45-72. Vitamin D and calcium were also included in the study, which showed that neither calcium nor vitamin D had any effect on prostate cancer and are not the cause. The first study was over a course of eight years. A similar study from University of Hawaii scientists tested the food groups and consumption of dairy products. They found their link—the consumption of low-fat or nonfat milk increased the risk of localized tumors or non-aggressive tumors. The same study showed that drinking whole or raw milk decreased this risk.

There is little data that validates the claim that low-fat milk like 1-2% and skim is healthy. Instead, it was promoted by the media to help lower calories in our diet. There is substantial evidence that low-fat foods and drinks don't give you satisfaction because they are less filling, and you end up consuming more. In a study published in the *Archives of Disease in Childhood* in March 2012, scientists found that children who drank lower-fat milk were actually more likely to be overweight later on.

"Our original hypothesis was that children who drank high-fat milk, either whole milk or 2%, would be heavier because they were consuming more saturated-fat calories. We were really surprised when we looked at the data and it was very clear that within every ethnicity and every socioeconomic strata, that it was actually the opposite, that children who drank skim milk and 1% were heavier than those who drank 2% and whole," study author Dr. Mark Daniel DeBoer, an associate professor of pediatric endocrinology at the University of Virginia School of Medicine and the chair-elect for the AAP Committee on Nutrition, told TIME in March.

People don't know that low-fat products are high on the glycemic index. This in turn can increase the level of triglycerides that can heighten the effect of heart disease while increasing the risks of high cholesterol and high blood pressure.

Pasteurization is the most widely used form of treating milk by bringing it to a high temperature. The healthy bacteria and enzymes are killed during this process, which gives the milk a longer shelf life. Not only does pasteurization kill the enzymes, but it reduces the vitamin content, destroys vitamins C, B12, and B6, kills beneficial bacteria, promotes pathogens, and is linked with allergies, increased tooth decay, colic in babies, growth problems in children, osteoporosis, arthritis, nerve disease, and cancer.

During the process of making pasteurized milk, slime must be removed, along with pus. Pus and slim are in the milk naturally and is increased with growth hormones and antibiotics. This is accomplished by a process of centrifugal clarification. One of the main reasons that pasteurization started back in the 1920s was to reduce the spread of TB, infant diarrhea, and undulant fever. These were prevalent primarily because the very poor industrial standards in the 1920s.

Not all bacteria are killed with the pasteurization method. One is the bacteria for Johne's disease (bacterial disease of the intestinal tract, Johne's disease is caused by Mycobacterium paratuberculosis. This bacterium is a relative of the bacterium that causes tuberculosis in humans).

Many people believe, is the cause of Crohn's disease in humans. Most commercial milk is now ultra-pasteurized to get rid of heat-resistant bacteria and give it a longer shelf life. Ultra-pasteurization is a process that takes milk from a chilled temperature to above the boiling point in less than two seconds.

Sterilized milk and UHT are basically the combination of two separate processes, homogenization and pasteurization. This form of milk is available in whole, semi-skimmed, and skimmed varieties. It goes through a more severe form of heat treatment, which destroys nearly all the bacteria enzymes. That's why it can last a half a year unopened on the shelves without needing to be refrigerated. After the container is opened, it needs to be consumed within five days. Because this is a combination of two separate forms of milk treatment, it encompasses the dangers and loss of nutrients in both.

Evaporated milk is a concentrated, sterilized milk product. The process of creating evaporated milk involves standardizing, heat treating, and evaporating the milk under reduced pressure at temperatures between 140°F and 149.°F. Evaporated milk is also homogenized to prevent it from separating in storage, and then it is cooled.

Condensed milk is made using the same process as evaporated milk. The only difference is the addition of sugar. Because of the high amounts of sugar, condensed milk is not sterilized. After the heating of the milk, it is then homogenized. The sugar is added, and then the milk is reduced in volume. The concentration of the condensed milk is now up to three times that of the original milk. As such, the uncooked rBGH may still remain in very concentrated amounts.

Filtered milk goes through a very fine filtration system which prevents enzymes and bacteria from passing through. The nutritional content of the milk is slightly affected, and the shelf life is increased. The processes for making this form of milk include microfiltration, ultrafiltration, and nanofiltration. This milk has been homogenized and pasteurized. Filtered milk is available in whole, semi-skimmed, or skimmed milk varieties.

Dried milk powder contains oxidized cholesterol, which contributes to the buildup of plaque in the arteries, leading to atherosclerosis. When you consume products that contain dried or powdered milk, with the understanding from the media that it will help you avoid heart disease, you're truly consuming oxidized cholesterol, which can lead to heart disease.

What is Oxidized Cholesterol?

Oxidized cholesterol is particularly irritating to your blood vessels, the chemical process of turning milk into a power oxidizes the cholesterol, irritation is what prompts the formation of plaque, which is the beginning of heart disease. Oxidized cholesterol molecules can in turn oxidize other cholesterol molecules, setting forth a sort of chain reaction.

Goat's milk makes up about half the world's milk consumption, even though not many Americans drink it. Goat's milk is less likely to cause an allergic reaction, which is why people with lactose intolerance are able to consume it with ease.

As goat's milk is naturally homogenized, the cream won't separate. It's closer to human milk and easily digested by infants. Many people who can't breastfeed use goat's milk as a base for their infant formula.

According to the *Journal of American Medicine,* **"Goat's milk is the most complete food known."** It contains vitamins, minerals, electrolytes, trace elements, enzymes, protein, and fatty acids that are utilized by your body with ease. In fact, your body can digest goat's milk in just twenty minutes. It takes 2-3 hours to digest cow's milk.

Excerpt from **"The Maker's Diet"** by Jordan S. Rubin Beneath are some of the health benefits attributed to raw goat milk consumption:

• Goat's milk is less allergic - It doesn't contain the complex proteins that stimulate allergic reactions to cow's milk.

• Goat's milk doesn't suppress the immune system.

• Goat's milk has a more buffering capacity than over the counter antacids. (The USDA and Prairie View A&M University in Texas have confirmed that goat's milk has more acid-buffering capacity than cow's milk, soy infant formula, and nonprescription antacid drugs.)

• Goat's milk alkalinizes the digestive system. It actually contains an alkaline ash, and it doesn't produce acid in the intestinal system. Goat's milk helps to increase the pH of the blood stream because it is the dairy product highest in the amino acid L-glutamine. L-glutamine is an alkalinizing amino acid often recommended by nutritionists. Pg. 148 - "The Makers Diet"

• Goat's milk contains twice the healthful medium-chain fatty acids, such as cupric and caprylic acids, which are highly antimicrobial. (They actually killed the bacteria used to test for the presence of antibiotics in cow's milk!)

• Goat's milk doesn't produce mucus; it doesn't stimulate a defense response from the human immune system.

• Goat's milk is a rich source of the trace mineral selenium, a necessary nutrient, is known for its immune modulation and antioxidant properties. Pg 149 "The Maker's Diet"

Easier digestion allows the lactose to pass through the intestines more rapidly, not giving it time to ferment or cause an osmotic imbalance. Goat's milk also contains 7% less lactose than cow milk. Additionally, most lactose intolerant people have found that they can tolerate goat's milk and goat milk products. Natural breast milk contains many bioactive components, which serve to retard the growth of harmful organisms, and to protect the health of the person consuming them. Goat's milk contains the same important bioactive components as mother's milk.

Medicinal properties of goat's milk: the importance of feeding of infants with goat milk has been recognized since ancient days. In countries like the U.S, the goat's milk is specifically marketed for the infants. The milk allergy problem common in infants fed with cow milk is rarely encountered when replaced with goat's milk, and it plays an important role in the making of infant formula. This is proof of the medicinal property of goat's milk. Symptoms like gastrointestinal disturbances, vomiting, colic, diarrhea, constipation, and respiratory problems can be eliminated when goat milk is fed to infants.

The relief in respiratory problems when fed with goat's milk can be attributed to the structure of the casein micelle of the goat's milk. Pasteurized goat's milk is well tolerated by infants with gastrointestinal or respiratory symptoms. Fermented goat's milk products are ideal for persons allergic to cow milk. The goat's milk is naturally homogenized. It forms a soft curd when compared to cow's milk, and hence helps in easy digestion and absorption. Regular intake of goat's milk significantly improves body weight gain, improved mineralization of the skeleton, and increased blood serum vitamin, mineral, and hemoglobin levels. These points are considered advantageous when compared to the consumption of human milk.

The other medicinal property of goat's milk is a higher concentration of medium chain fatty acids, which play an important role in imparting unique health benefits in malabsorption syndrome (small intestine cannot absorb nutrients from foods), steatorrhea (excretion of abnormal quantities of fat with the feces owing to reduced absorption of fat by the intestine), chyluria, hyperlipoproteinemia, and during conditions of cystic fibrosis, gallstones, and childhood epilepsy. The medium chain fatty acids minimize cholesterol deposition in the arteries, aid in dissolving cholesterol and gallstones, and significantly contribute to the normal growth of infants.

Soy milk has less protein than dairy milk, and contains "anti-nutrients" that actually decrease the absorption of calcium, magnesium, copper, iron, and zinc. Soy milk prohibits the absorption of vitamins. The isoflavones that are in soy products are phytoestrogens, which are estrogens that act like human hormones. Soy products contain enough isoflavones to cause severe disruption to the hormonal systems of infants during a critical period in their development. There are studies showing that soy is playing a role in the current epidemic of infertility, menstrual, and other reproductive problems in humans.

Rice milk has no saturated fat or cholesterol, and barely any protein. The FDA has not required arsenic reduction as part of the processing of rice-based products.

Almond milk is a goitrogenic food that comes from the almond nut. This means that almond milk has chemicals that harms the thyroid. This kind of food expands the thyroid. Excess consumption of these goitrogenic foods are known to cause goiters, and stop the body from getting the sufficient absorption of iodine by the body. If you have good iodine intake, small amounts of almond milk would be fine, but I wouldn't recommend consuming large amounts of goitrogenic food without having sufficient iodine in your diet. This is just a note—some forms of almond milk contain salmonella, which is why they pasteurize all forms of almond milk, even though no organic almonds have been found with salmonella.

Hemp milk has given rise to a lot of confusion. First off, hemp is not marijuana, even though hemp and marijuana are both from the same cannabis family of plants. Hemp has been used for thousands of years in treating everything from cancer to toothaches. Hemp milk is starting to become very popular in vegan and vegetarian diets because of the health benefits and high nutritional value. Hemp milk is not only rich in fat, unsaturated fats, which are frequently called the "good" fats. A single glass of hemp milk contains 900mg of omega-3 fats and 2800 mg of omega-6 fats. These fats are within the 3:1 ratio of the omega-6 to omega-3 that diet experts consider ideal. Nutrient Breakdown for Hemp Milk (per liter):

- 900mg omega-3

- 2800mg omega-6

- 10 essential amino acids

- 4 grams of healthy protein

- Vitamin A, D, E, B12, riboflavin, and folic acid

- Magnesium, phosphorus, zinc, and iron

- 46% of Minimum recommended dietary allowance (RDA) of calcium

With all the forms of the milk products above, research shows that the organic form of milk has far more benefits than the GMO counterpart. Many of the forms of milk are available in organic or non-GMO.

Section 18
The Truth About BPA

Bisphenol A, more commonly known as BPA, is a manufacturing chemical that has been used in the production of certain plastics and resins since the 1960s. BPA is commonly used in polycarbonate plastics and epoxy resins. These plastics are used for the production of food and beverage containers, such as baby and water bottles. The resin commonly used in the lining of tin can products, such as water supply lines, tuna cans, some dental sealants, and mixtures also may be composed of BPA.

Unless you have been living under a rock, you've heard about BPA and seen BPA-free labeled products, even though the FDA claims that the use of BPA is safe in low doses. Many people don't know why BPA is so bad for your health. It's a man made chemical that mimics estrogen ingested into the body. Numerous health experts and consumers are concerned over the long-term health effects at both small and high doses of BPA. A study was conducted in England at the National Health and Nutrition Examination Surveys back on January 25, 2010, have found a shocking discovery. They found when males with high levels of BPA exposure had a 10% increase in the risk of developing heart disease.

A group of French researchers published findings that links BPA and negative effects on the intestines. The intestines are one of the first organs that comes in contact with BPA after its consumption. In animal studies, the researchers used levels that were ten times below the levels that the FDA claims are safe for humans. The mucosal lining of the intestinal wall failed, leading to a condition called "leaky gut syndrome." This condition can cause toxins and bacteria to leak into the body and damage tissues and organs.

Animal study indicated that exposure to even low levels of BPA can affect the function of female antral follicles, the egg cell which produces the egg during ovulation. It tricks the body, binding with the estrogen receptors in the cells. BPA is linked to lowering the progesterone levels, a vital hormone that's responsible for the female reproductive system and is connected with the female menstrual cycle, pregnancy, and embryogenesis (which is the process of the formation of the embryo and its development).

Study was conducted in China focused their efforts on the use of BPA in men. The journal Human Reproduction published a research study that was conducted with over 200 men who were exposed to BPA at their place of work. These men were four times more likely to have erectile dysfunction, including ejaculation difficulties. Even though the levels of BPA exposure were fifty times higher than any American would come in contact with, the study showed that BPA has a negative effect on the male reproductive organs. The study did not look at long-term exposure to BPA and male hormones at low levels over a period of time. In my opinion, the reason that erectile dysfunction, including ejaculation difficulties are becoming more common is estrogen mimickers such as BPA. Some studies state that BPA at the level are not harmful and do not have an effect—I just think that the effect would be called "normal," and would be dismissed.

Research has shown that BPA contributes to diabetes and metabolic syndrome. These two conditions are caused by the body not effectively utilizing insulin, also known as insulin resistance. BPA is linked to the body's overproduction of insulin and over use of the pancreas. In my opinion, BPA could also be linked to polycystic ovarian syndrome (PCOS), as both type 2 diabetes and polycystic ovarian syndrome are related to overproduction of insulin, as the increasing levels of insulin within the body prevents the body's ability to metabolize fat, which could lead to obesity.

A University of Chapel Hill study has shown the danger of BPA for young children exposed to BPA from baby bottles. Young girls who were exposed to BPA showed higher aggressive and hyperactive tendencies than children with no or very little exposure to BPA. BPA exposure in mothers can travel from the umbilical cord to the infant.

The graph below is by the Environmental Working Group and shows the levels of BPA they found in several foods.

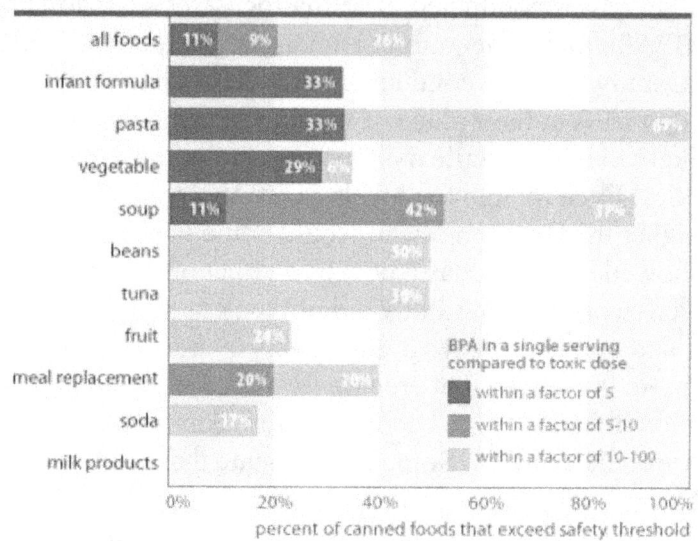

BPA IS AT UNSAFE LEVELS IN ONE OF EVERY 10 SERVINGS OF CANNED FOODS (11%) AND ONE OF EVERY 3 CANS OF INFANT FORMULA (33%)

Source: Chemical analyses of 97 canned foods by Southern Testing and Research Division of Microbac Laboratories, Inc., North Carolina.

"The tests showed that single servings of almost half of the products had levels of BPA comparable to levels that laboratory studies have linked to adverse health effects," reported Shannon Coughlin of the Breast Cancer Fund. **"Then there's the leftovers. And then there's the rest of the year— most of us eat canned foods daily or weekly. Even if we don't use them much at home, canned foods are used in a lot of restaurant and cafeteria food. When you think of this daily**

exposure, you start to see the urgency of getting this chemical out of food cans."

BPA exposure can increase some cancer cells, these cells may become resistant to chemotherapy and harder to treat, according to a study from Duke University. Duke focuses on inflammatory breast cancer, which accounts for 1 out of 5 of all breast cancer cases that are diagnosed, BPA might have made the treatment insignificant. The same research even suggests that BPA might even be the underlying cause of the cancer cell growth.

In many animal studies, estrogen initiates or promotes cancer of the prostate. Some scientists speculate that estrogen and estrogen mimickers, such as BPA, may be one of the causes of the increase of prostate cancer. A study by Gail Prins at the University of Illinois at Chicago Department of Urology, Shuk-Mei Ho of the University of Cincinnati Department of Environmental Health, and their colleagues provides the first evidence of a direct link between low-dose BPA exposure during development and later prostate cancer. In modern America, the false hormone produced by BPA and estrogen mimickers alike is very common, and can be detected in 90% of the blood samples that was investigated, by Prins' estimate. According to Prins' research, the BPA levels were measured from the tissue of the placenta and the fetus, and the chemical may be five times higher than in the blood. The levels are higher in males than females. If BPA is so bad for you, is the alternative any better? Some independent research has shown that BPA-free-labeled containers had estrogen mimickers in them, and some at even higher amounts than BPA. Consumers have been buying BPA-free products, thinking they were safer.

Always use glass and stainless-steel containers, you can limit the contact with estrogen-like chemicals. According the research conducted by Toxicological Sciences, BPA replacements such as bisphenol-S (BPS) or bisphenol-F (BPF) have similar effects and might actually be even worse than BPA. **"The chemicals have the same function [as BPA], which usually means they're similar in structure, and therefore have similar**

health effects," says Lindsay Dahl, deputy director of the organization Safer Chemicals, Healthy Families. These chemicals leak from the plastic into our bodies by common practices such as heating or cooling the plastic.

The worst part is that there is no federal law that forces companies to test their chemicals to see if they are proven safe before they are used for public use. **"So if a manufacturer decides to stop using BPA, they have no laws to follow that require them to use a safer chemical. As a result, they've been switching to chemicals that work the same,"** Dahl says.

Section 19
Bottled Water

The media would have you believe that bottled water is healthier than normal tap water, but is this true? There is a lot of conflicting information, I would like to clear up that confusion and give you some clarity on the truth. There has been some speculation that the extraction of water for the production of bottled water may be contributing to the drought in California. Below is a map from *Mother Jones*. They created this illustration using the location of bottled water companies such as Dasani, Aquafina, Crystal Geyser, and Arrowhead. They overlaid that data on the current U.S. Drought Monitor, the measurement of the drought all over the country.

Regional Water-quality Assessments of Principal Aquifers
click on an aquifer name to learn more

This map shows the use of water may be affecting the environment and creating man made water shortages all over the United States. This could be happening all over the world where water extraction were being performed. The two main ways that water extracted for bottling are by means of drilling an aquifer and draining the water with large pumps, while about 45 percent of it is just treated tap water.

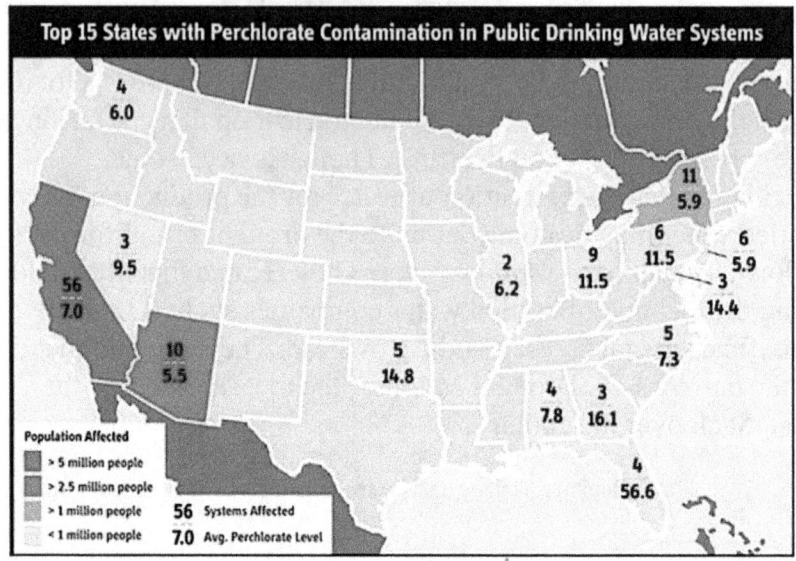

Map from water.usgs.gov.

This map shows the locations of the aquifers within the United States. The locations of the aquifers match the data for the drought locations in the United States. Could extracting the water from the ground be part of the water shortage that the western half of the United States is experiencing? Some research is saying YES!

Posted 2014-08-25 by The Inquisitr: **"Based upon a study published in *Nature*, geologists believe it's possible that a lack of water in the San Joaquin Valley is decreasing the weight on the San Andreas Fault, which has led some to make predictions of more earthquakes in the San Francisco Bay area. In 2007, a panel of experts estimated there was a 63 percent chance that the Bay area will experience another catastrophic earthquake in the next 23 years. Due to there being less water in the aquifer, the soil in the area is sinking, causing the land to drop about a foot a year."**

Research is showing that it is very possible that water excavation may have a large part, water excavation may be affecting the environment and climate all over the United States. In the earlier section, we talked about the dangers of BPA and BPS. Drinking water from plastic bottles is very dangerous to your health.

BPA - Bisphenol A or BPA is a chemical that has been linked to a multitude of serious health problems, including:

- Learning and behavioral problems

- Altered immune system function

- Early puberty in girls and fertility problems

- Decreased sperm count

- Prostate and breast cancer

- Diabetes and obesity

Phthalates: Phthalates are used in the United States to make plastics like more flexible.

Phthalates are chemicals linked to developmental and reproductive effects such as:

- Reduced sperm counts

- Testicular atrophy or structural abnormality

- Liver cancer

Research conducted on phthalates has shown males are getting their fetal androgens blocked before birth, which may very well affect gender development in male offspring, and could lead to undescended testes at birth and testicular tumors later in life. Phthalates exposure in pregnant mothers makes it more likely to have baby boys who have certain effeminate traits and produce less testosterone, and might be born up to a week early. You're not just exposed to BPA in bottled water—you could also be exposed to fluoride, chlorine, arsenic, aluminum, uranium, disinfection byproducts, and prescription drugs.

Published by *Ban The Bottle* on November 17, 2010 by Tomás Bosque:

According to the FDA: **"Bottled water is water that's intended for human consumption and that's sealed in bottles or other containers with no added ingredients except that it may optionally contain safe and suitable antimicrobial agents. Fluoride may be optionally added within the limitations established."**

The FDA regulates bottled water, which is classified as a "food." The Environmental Protection Agency regulates tap water. The Environmental Protection Agency has stricter rules for tap water than that of the FDA for bottled drinking water. This means that the water in your shower in might be safer than the water that you're drinking out of that bottle. The FDA's specific regulations for bottled water are found in Title 21 of the Code of Federal Regulations (21 CFR). Under the standard of quality (21 CFR, 165.110 [b]), the FDA allows certain levels of contaminants in bottled water.

Some of the contaminants are:

1. Coliforms are bacteria that are always present in the digestive tracts of animals, including humans, and are found in their wastes. The FDA will allow up to 9.2 coliform organisms per 100 milliliters.

2. Arsenic and its compounds, especially trioxide, are used in the production of pesticides, treated wood products, herbicides, and insecticides. The FDA says that bottled water may have up to 0.05 milligrams per liter of arsenic.

3. Chloride is a compound of chlorine, a substance used to disinfect tap water. The FDA allows up to 250.0 milligrams per liter of chloride in bottled water.

4. Iron is a metallic element. Your body needs some iron. The FDA permits bottled water to contain up to 0.3 milligrams per liter of iron.

5. Manganese is used in glass making and in fertilizers. Bottled water may contain up to 0.05 milligrams per liter of manganese.

6. Phenols are corrosive, poisonous acidic compounds, also known as carbolic acid. Bottled water may contain up to 0.001 milligrams per liter of phenols.

7.Dissolved solids, comprise inorganic salts (principally calcium, magnesium, potassium, sodium, bicarbonates, chlorides, and sulfates) and some small amounts of organic matter that are dissolved in water. The FDA allows bottled water to contain up to 500 milligrams per liter of dissolved solids of whatever type.

8. Zinc is a metallic element. Your body needs zinc. The FDA permits bottled water to contain up to 5.0 milligrams per liter of zinc.

9. Fluoride is purposely added to some bottled water. Bottled water that's not labeled as containing fluoride may contain up to 2.4 milligrams per liter of fluoride.

Chemical contaminants:

The FDA allows some shocking levels of very toxic chemicals in bottled water. Here is a sampling of chemical contaminants bottled water has in it, along with the permitted milligrams per two gallons. The amounts may vary.

- Barium.. 2.0
- Chromium.. 0.1
- Copper.. 1.0
- Cyanide... 0.2
- Nickel... 0.1

- Ethylbenzene (100-41-4)..................... 0.7

- Monochlorobenzene (108-90-7)............... 0.1

- Styrene (100-42-5)............................ 0.1

- Toluene (108-88-3)............................ 1.0

- Xylenes (1330-20-7)......................... 10.0

- Uranium......... 30 mg/L 180mg per two gallons

According to the *EPA*:

"Exposure to uranium in drinking water may result in toxic effects to the kidney. Some people who drink water containing alpha emitters, such as uranium, in excess of the MCL over many years may have an increased risk of getting cancer." The FDA allows a certain amount of pesticides in bottled water, there are set limits for each of twenty-nine different pesticides. There have been disinfectants found in bottled water, as the FDA allows this water to contain set levels of disinfectants and disinfection byproducts. Below is a list of what might be in your bottled water.

Examples from 21 CFR (Code of Federal Regulations) 165.110:

Disinfection byproducts

- Bromate...
...... 0.010

- Chlorite...
.... 1.0

- Haloacetic acids (five)
(HAA5)....................................... 0.060

- Total Trihalomethanes
(TTHM)....................................... 0.080

Residual disinfectants

- Chloramine.. 4.0 (as Cl2)

- Chlorine... 4.0 (as Cl2)

- Chlorine dioxide................................. 0.8 (as ClO2)

The FDA allows bottled water to contain certain amounts of radioactive material. See 21 CFR (Code of Federal Regulations) 165.110.

Three examples:

- "The bottled water shall not contain a combined radium-226 and radium-228 activity in excess of 5 picocuries per liter of water."

- "The bottled water shall not contain a gross alpha particle activity in excess of 15 picocuries per liter of water."

- "The bottled water shall not contain uranium in excess of 30 micrograms per liter of water."

Bottled water has in it more than regulations allow. When bottled water doesn't meet the standards set out by the FDA, it might still be sold. By law, it should bear a suitable label.

Examples:

1. "Contains Excessive Bacteria"

2. "Contains Excessive Arsenic"

3. "Excessively Radioactive"

So-called vitamin water is one of the biggest trends that's happening right now. In truth, vitamin water is the absolute worst kind of bottled water on the market. Most of it contains additives such as high fructose corn syrup and food dyes that can wreak havoc on your physical and emotional health.

Section 20 Non-eatables

When people talk about healthy shopping, they seem to miss the non-eatables section. This section includes the things that people use every day, but don't think about their side effects. Some of these products can be just as damaging to your health as their food counterparts. These toxins can be absorbed by your skin and through breathing them in, seeping into your bloodstream and being stored in your fat tissues. This can lead to cancer, changing DNA, causing infertility, neurotoxicity, or an array of other medical problems. This change will take time—there lies the danger. You don't even know about it until it's too late.

The cosmetics industry is a multibillion-dollar industry. What many people don't know is they use very toxic chemicals which were shown in animal studies to be linked to birth defects, cancer, and skin conditions. These chemicals are found in things that look harmless, like shampoos, conditioners, antiperspirants, cosmetics, and especially nail polishes.

A report published by the Environmental Working Group stated that they found some disturbing evidence that showed a very toxic chemical was found in every single person who was tested. This chemical is called dibutyl phthalate, or DBP. The CDC found in their research that the absolute highest levels were in women of reproductive age. **"Major loopholes in federal law allow cosmetics manufacturers to put unlimited amounts of industrial chemicals like DBP into personal care products without any testing for adverse health effects,"** the reports says.

The absolutely crazy thing is that there is no government law that forces the cosmetic industry to conduct chemical safety tests. The only law is if the chemical is added to the food. DBP is linked to male reproductive effects that come from the mother using it during pregnancy.

Research from animal studies shows that the chemical DBP lowers the testosterone levels in developing unborn males. This could cause the male genitalia to change. This can include undescended testicles, hypospadias, and sexual ambiguity. The EPA conducted studies on DBP chemical, and noted that the effect in the lab studies may just be applicable to humans. Have you ever noticed that male enhancement pills are now becoming more popular among younger men?

Epidemiological studies have linked the use of phthalate exposure to the drop in testosterone levels and increased sexual ambiguity. This is even evident in newborn males. Even though many companies are starting to hear the voices of their consumers about the dangers of DBP and are starting to take it out of their products, there are still many companies that are not, and it is still in use to this day. There are many kinds of phthalates, including DEHP, DBP, DINP, DIDP, and BBP. Each has their own specific uses and properties, and they each affect the body differently.

DBP is used in a wide range of products, such as nail polish, shampoos, conditioners, lotions, hair growth formulations, antiperspirants, sunscreen, even gum and candy. It doesn't end with DBP—there are numerous chemicals in cosmetics, and some of the components that are used as thickening agents can cause some outrageous health problems.

In 2012, California listed Cocamide DEA as a known carcinogen for causing cancer under its Prop 65 law. This law requires warning labels on consumer products containing carcinogens or reproductive toxicants that can cause cancer, allergies, immunotoxicity, and concerns over organ system toxicity. In 2013, the Center for Environmental Health filed a California lawsuit against companies that violated the law that was established in 2012 and sold shampoos and personal care products containing the toxic chemical without a warning label.

These products were sold in stores like Walmart, Target, Babies 'R' Us, and many more major retailers nationwide. Some name brands contained the carcinogen, as did store brands sold at

Walmart, Trader Joe's, Pharmaca, and Kohl's, according to the nonprofit. **"Most people believe that products sold in major stores are tested for safety, but consumers need to know that they could be doused with a cancer-causing chemical every time they shower or shampoo,"** said Michael Green, executive director of the Center for Environmental Health. **"We expect companies to take swift action to end this unnecessary risk to our children's and families' health."**

In this case, not even organic is safe. Deceptively labeled organic products from Africa's Best also tested for high levels of DEA. This shows you that you MUST READ THE LABEL, you can gain a better understanding of what is in the product that you are using.

MEA is an organic chemical, it acts as a surface active agent, emulsifier, and plasticizing agent. It's often applied as a pH balancer in products. MEA can be damaging when you breathe it in and apply it on your skin. It's a known human immune system toxin; a very corrosive chemical. Contact can severely irritate and burn the skin and eyes, with the potential of long-term eye damage. It may damage the liver and kidneys, and it may affect the nervous system.

Used for: facial moisturizer/treatment, shampoo.
listed on labels as: mea; monoethanolamine.

TEA- Triethanolamine—this is often used as a pH balancer. You will often find it in hypoallergenic products. It's linked to causing allergic reactions, including eye problems, dryness of hair and skin, and could be toxic if absorbed by the skin over a long period of time. TEA is restricted in the UK, as it has a well-known carcinogenic effect. However, it's still in use in the United States.

Synonym (s): triethanolamine; trolamine; tea; triathanolamine; thriethanolamine; treithanolamine; tritethanolamine; trietnanolamine; triethanlolamine; trethanolamine; triethanlamine; trethamolamine; trietahnolamine; thiethanolamine.

PEG or ethoxylated alkyloamides—polyethylene glycol (PEG). PEG-6, PEG-150 and other similar names are all members of the PEG family. The PEG chemical compounds are synthetic chemicals used in cosmetics as surfactants, cleansing agents, emulsifiers, skin conditioners, and humectants. The research shows links to an increased risk of a variety of cancers, including breast cancer. They contain various harmful toxins, according to a report by the Cosmetic Ingredient Review (CIR) committee that was published in the International Journal of Toxicology, including:

> • Ethylene oxide: Ethylene oxide increases the incidences of uterine and breast cancers and of leukemia and brain cancer, according to experimental results reported by the National Toxicology Program.

> • 1,4-dioxane: According to the National Toxicology Program, "1,4-dioxane is reasonably anticipated to be a human carcinogen."

> • Polycyclic aromatic compounds (PAHs): Known to increase the risk of breast cancer.

According to the Environmental Working Group, the following percentages of common cosmetics and hair products contain PEG compounds.

Mousse 90.3%

Hair Dye 79.5%

Baby Bath Wash 73.8%

Douche/Personal Cleanser 58.3%

Menopause Cream 54.5%

Depilatory Cream/Hair Remover 48.2%

Baby Lotion/Oil 46.4%

Anti-Itch/Rash Cream 46.3%

After-Sun Products 45.5%

Lip Balm/Treatment 43.6%

Moisturizer 43.1%

Deodorant 42.7%

Facial Moisturizer/Treatment 42.0%

Shaving Products 41.3%

Anti-Aging Treatment 41.0%

Styling Product 39.6%

Eye Treatment 38.8%

Concealer 37.9%

Foot Odor/Cream/Treatment 37.3%

Conditioner 35.2%

The other chemicals to look out for are: cocoyl sarcosine, 2-bromo-2-nitropropane-1,3-diol, imidazolidinyl urea, hydrolysed animal protein, lauryl sarcosine, quaternium-7, 15, 31, 60, etc, sodium lauryl sulfate, ammonium lauryl sulfate, sodium laureth sulfate, ammonium laureth sulfate, and sodium methyl cocoyl taurate.

These chemicals are commonly used as emulsifiers and foaming agents in the majority of body care products, even though the FDA (the American Food and Drug Administration) has cautioned the cosmetics industry of their potential danger since 1979. Back in 1998, a study by the NTP (American National Toxicology Program) protested against their use and labeled them nitrates and nitrosamines, cancer-causing elements.

An Environmental Working Group analysis shows that **"89 percent of 10,500 ingredients used in personal care products have not been evaluated for safety by the CIR (Cosmetic Ingredient Review), the FDA (Food and Drug Administration), nor any other publicly accountable institution."**

DEET- Is a plastic solvent used as an insect repellent. From the Duke University Medical Center News Office:

Posted on June 21, 2002 by Cabell Smith, Duke Today News.

"Every year, approximately one-third of Americans use insect repellents containing the insecticide DEET. Duke University Medical Center pharmacologist Mohamed Abou-Donia has spent 30 years researching the effects of pesticides. He has found that prolonged exposure to DEET can impair functioning in parts of the brain."

"Damage to these areas could result in problems with muscle coordination, muscle weakness, walking or even memory and cognition."

Abou-Donia says rats given even small doses of DEET for 60 days had a harder time accomplishing even the easiest tasks. Abou-Donia says short-term exposure to DEET doesn't appear to be harmful, but warns against using any product with more than a 30 percent concentration. Use as little of the product as you can, and don't use a product containing DEET if you're taking any medication. **"We found that the combined exposure to DEET and other chemicals is more dangerous than just DEET alone."** Abou-Donia also warns to never put a product containing DEET on an infant's or child's skin. The side effects could be even more serious.

Synonym (s): N, N-diethyl-meta-toluamide

Propylene Glycol- The Material Data Safety Sheet (MSDS: charter for use of the products) of propylene glycol cautions against contact with the skin. It could cause dermatitis, ototoxicity, kidney damage, and liver problems, according to various clinical studies. Commonly found in vaping/e cigarettes.

Sodium Lauryl Sulphate (SLS) and Sodium Laureth Sulphate (SLES)- A report published in the *Journal of the American College of Toxicology* back in 1983 stated that even low concentrations like 0.5% could cause irritation, and concentrations of 10-30% caused skin corrosion and severe irritation.

National Institutes of Health **"Household Products Directory," they stated that over 80% of products contain sodium lauryl sulfate. And many of them had concentrations up to 30%, which the ACT report called "highly irritating and dangerous."**

Other dangers of SLS are skin irritation, hormonal imbalance, eye problems in children, and cancer. Commonly used in soaps, shampoos, detergents, and toothpastes and other beauty products.

Stearalkonium Chloride- This chemical was developed as an ingredient for fabric softener. Stearalkonium Chloride is often found in hair conditioners and creams. It's a toxic chemical that can cause allergic reactions on contact.

"Although it's a proven irritant, many companies use it in hair conditioning products because it's cheaper and easier to incorporate than proteins or herbals," says Masters.

Function (s): antistatic agent; preservative; surfactant

Synoym(s): benzenemethanaminium, n,n-dimethyl-n-octadecyl-, chloride;
benzenemethanaminium, n,ndimethylnoctadecyl, chloride; benzyl dimethyl stearyl ammonium chloride; benzyldimethyl (octadecyl) ammonium chloride; chloride benzenemethanaminium, n,n-dimethyl-n-octadecyl-; n,n-dimethyl-n-octadecyl- chloride benzenemethanaminium; n,n-dimethyl-n-octadecylbenzenemethanaminium chloride; stearyl dimethyl benzyl ammonium chloride; 2b (onium compound) ; ammonyx 4; ammonyx 4002

Synthetic Perfumes- A massive 90% plus of the chemicals that make up a synthetic fragrance are derived from petrochemicals (petroleum or natural gas). These chemicals can actually make people go into anaphylactic shock, trigger a painful migraine, set off an asthma or allergy attack, dizziness, hyper-pigmentation, violent cough, vomiting, and skin irritation. They can cause any number of other symptoms, and contribute to chronic ill health.

"Sensitizers: One in every 50 people may suffer immune system damage from fragrance and become sensitized, according to the EU's Scientific Committee on Cosmetic Products and Non-food Products. Once sensitized to an ingredient, a person can remain so for a lifetime, enduring allergic reactions with every subsequent exposure."

Talc- (also known as talcum powder). There have been over forty year's worth of research studies conducted on the long-term effect of talc. Talc is used in products ranging from baby powders to makeup. This seemingly harmless ingredient that's used in many products can actually cause frightening side effects, among them **ovarian, lung, and other cancers.** Studies that were conducted before 2003 from sixteen independent researchers determined a 30% increase in the risk of ovarian cancer by people who used talcum powder. The research stated that women that use talcum powder for personal hygiene are particularly at risk.

Harvard epidemiologist claims that as many as 21,290 women are diagnosed with ovarian cancer every year. The research suggests that it is in part because of their use of baby powder. The talc that travels within the body and the female reproductive system could very well cause long-term inflammation of the ovaries. This in turn leads to cancer cell growth. These transfers of talc particles could be from any talc that's in contact with the genital area. This contact could be from sanitary and incontinence pads, diaphragms, body powders, and some condoms, for example. Ovarian cancer is only 3% of the cancer diagnosed, but is the deadliest of all the cancers that attack the reproductive system.

Talc is also damaging to your lungs, even though the FDA has claimed that cosmetic-grade talc is considered safe. At the National Toxicology Program (NTP) in Britain back in 1993, they had found that lab rats that had been exposed to cosmetic-grade talc by means of inhalation had developed a variety of inflammatory lung disorders, including cancer of the lungs and rare adrenal cancers.

According to the Federal Food, Drug, and Cosmetic Act (FD&C Act), any cosmetic products and their ingredients, with the only exclusion being that of color additives, don't have to be tested by the FDA before they hit the market shelves.

Aluminum- Aluminum compounds are known to be neurotoxic to humans. Did you know that aluminum is used in many antiperspirants and antiseptics? There have been numerous scientific studies that have linked the use of aluminum to breast cancer. Some research suggests that aluminum may even play a role in Alzheimer's. Avoid products containing aluminum altogether. Some lotions can contain aluminum. It can be listed on the label as Magnesium Aluminum Silicate or another aluminum compound.

Aminomethyl Propanol- There are reports of neurotoxicity, organ toxicity, endocrine disruption, and ecotoxicity from using this product. It may contain nitrosamine compounds. This is due to some form of nitrosating agents that may be present. If this is the case, it's considered a carcinogen, and may cause cancer. The big problem with Aminomethyl Propanol is that it is near impossible to know whether nitrosamine compounds are present. The only way to know is by testing the product for nitrosamine.

Butylated Hydroxyanisole (BHA) / Butylated hydroxytoluene (BHT)- Not only are these synthetic chemicals, but they are carcinogenic. Animal studies that were conducted showed that BHA and BHT can cause metabolic stress, depression, weight loss, damage to the liver, baldness, and fetal abnormalities. BHA and BHT are suggested to be one of the links that could be the causes of ADHD and behavioral disturbances in children. BHA & BHT both act as a synthetic estrogen or xenoestrogen. Up to 13% of the BHT is absorbed through the skin. Studies have shown that long-term exposure to BHT is toxic in mice and rats, causing liver, thyroid, and kidney problems and affecting lung function, blood thickening, and tumors. This could prevent lower male sex hormones, resulting in adverse reproductive effects. The research that has been conducted on BHT and BHA is very limited and ongoing. The evidence suggests that it could be a very big health concern.

Commonly found in: Food

Methylisothiazolinone and Methylchloroisothiazolinone- Both are found to be toxic. In countries around the world such as Japan and Canada, these chemicals and their counterparts are all but banned because studies have shown them to be killing the brain cells in rats. These chemicals are part of a class of chemicals called biocides. They are used in many common household items such as Head and Shoulders, Suave, and Clairol, as well as Pantene hair conditioner and Revlon hair color.

Synonym (s): 2-Methyl-4-isothiazolin-3-one, 5-Chloro-2-methyl-4-isothiazolin-3-one, Acticide, Algucid, Amerstat 250, Euxyl K 100, Fennosan IT 21, Grotan TK2, Kathon, Kathon CG, Mergal K7, Metatin GT, Methylchloroisothiazolinone, Methylisothiazolinone (MI), Mitco CC 32

Products that may contain Methylisothiazolinone and Methylchloroisothiazolinone mix:

Cosmetics

- Foundations/concealers
- Bronzers/self-tanners
- Eye shadows
- Mascaras
- Makeup removers
- Moisturizers

Pharmaceutical/self-hygiene products

- Sunscreens
- Shampoos/conditioners
- Bubble baths
- Soaps
- Baby wipes
- Creams/lotions/gel
- OTC and prescription medicines

Household/industrial products

- Detergents/cleaners
- Fabric softeners
- Polishes
- Pesticides
- Paints
- Adhesives/glues
- Latex emulsions
- Radiography
- Curing agents
- Jet fuels
- Printing inks

Polyacrylamide- This ingredient is not toxic, but when ingested, it breaks down and turns into acrylamide, which is a carcinogen. The research that has been conducted on lab rats have found that acrylamide do pose a risk of causing several kinds of cancer. The research on the effects on humans has not yet been completed. The National Toxicology Program and the International Agency for Research on Cancer consider acrylamide to be a **"probable human carcinogen."** This conclusion is based on the data collected by the laboratory studies on animals. Even though the human studies are not yet complete, I would still not use any product that contains any polyacrylamide.

Sodium Hydroxide- This is also known as caustic soda and is used in the plumbing industry in cleaning clogged drains. Caustic soda is also used in many oven cleaners. Sodium hydroxide is an ingredient in many toothpastes as a pH stabilizer and skin creams. With contact on skin, this could cause contact dermatitis.

Information direct from the Sodium Hydroxide Material Safety Data Sheet (MSDS): **"POISON! DANGER! CORROSIVE. MAY BE FATAL IF SWALLOWED. HARMFUL IF INHALED. CAUSES BURNS TO ANY AREA OF CONTACT. REACTS WITH WATER, ACIDS AND OTHER MATERIALS.**

> **Ingestion: Corrosive! Swallowing may cause severe burns of mouth, throat, and stomach. Severe scarring of tissue and death may result. Symptoms may include bleeding, vomiting, diarrhea, fall in blood pressure. Damage may appear days after exposure.**

> **Ingestion: Corrosive! Swallowing may cause severe burns of mouth, throat, and stomach. Severe scarring of tissue and death may result. Symptoms may include bleeding, vomiting, diarrhea, fall in blood pressure. Damage may appear days after exposure.**

> **Skin Contact: Corrosive!** Contact with skin can cause irritation or severe burns and scarring with greater exposures.

> **Eye Contact: Corrosive!** Causes irritation of eyes, and with greater exposure, it can cause burns that may result in permanent impairment of vision, even blindness.

> **Chronic Exposure:** Prolonged contact with dilute solutions, has a destructive effect upon tissue.

Aggravation of Preexisting Conditions: Persons with pre-existing skin disorders or eye problems or impaired respiratory function may be more susceptible to the effects of the substance.

According to the US agency that regulates cosmetics—the FDA's Office of Cosmetics and Colors: **"A cosmetic manufacturer may use almost any raw material as a cosmetic ingredient and market the product without an approval from the FDA. Testing of product ingredients is not only controlled by the manufacturers, but is also voluntary. Not surprisingly, then, many ingredients in cosmetics, skincare products, and other personal-care products are not tested for safety at all, and most have not been evaluated for safety by the FDA. This means that companies can market ingredients that are known to pose potentially serious health risks, including some found in moisturizers."**

Would you really buy a product if the label read, **"Aminomethyl Propanol. This ingredient may cause cancer."** I don't think anyone would. I believe that the companies use the ingredients in full consciousness, knowing that they cause harm. They claim that they are safe—however, the studies state otherwise. So why do the manufacturers still use them? It's simple. We as consumers still buy them.

Manufacturers are not required to notify their consumers that their product may cause long-term damage and even cancer. They are not required to say that it contains hormones from GMO animals, if there are animal ingredients in the product, nor do they change the ingredients to something that's not toxic. Some of the ingredients are even known to modify the genetics of the user and their offspring, but why do people not know this? That's because people don't understand the ingredients on the label—that's done by design.

Preservatives used in food and skincare products are to prevent spoilage before use, but many of the preservatives are linked to lifelong health side effects. The preservatives break down into a chemical called formaldehyde. In 2011, the National Toxicology Program named formaldehyde as a known human carcinogen in their 12th Report on Carcinogens. Here, we'll talk about the chemicals that are known to contain or break down into formaldehyde. There are many products that may contain this deadly toxin. They range from health and beauty supplies to food, cleaning, and home/furnishing products. Here is a list of some of the personal products that might contain this toxin:

Lotions	Shampoos	Sunblock
Soap	Cosmetics	Body wash
Toothpaste	Baby wipes	Baby wipes

I'm sure that while reading this book, many of you have been checking the labels of products before you buy them, and you're thinking, "I haven't seen any products that list formaldehyde on the label." Most manufacturers will use synonyms instead. Look out for products that contain the following ingredients:

Formalin	Methanal	Oxymethylene
Urea	1,3-Dioxetane	Quaternium 15
Methyl Aldehyde	Methylene Oxide	Formic Aldehyde

| Oxomethane | Phenol |
| Formalin | Formaldehy de |

You can look up a reference list on the U.S. Department of Health and Human Safety website. After I looked at the list, I was shocked at the countless infant and child products that contain this deadly chemical.

Sodium Hydroxymethylglycinate- Even though sodium hydroxymethylglycinate is often marked as a natural and safe preservative, this is not entirely true. The process for making it requires converting glycerin into sodium hydroxymethylglycinate, which is by no means a natural process. Sodium hydroxymethylglycinate is a relatively common ingredient in skincare products. According to the American Academy of Dermatology, this could be one of the main causes of eczema and contact dermatitis in skincare product by a preservative.

Function (s): Hair Conditioning Agent; Preservative

Commonly found in: Hair care, Moisturizer, Anti-aging, Facial cleanser, Body wash, Facial mask, Styling gel, Eye cream.

Synonym (s): sodium hydroxymethylglycinate, glycine, n- (hydroxymethyl) -, monosodium salt; glycine, n- (hydroxymethyl) -, sodium salt; glycine, n- (hydroxymethyl)-. monosodium salt; hydroxymethylaminoacetic acid, sodium salt; n- (hydroxymethyl) - sodium salt glycine; n- (hydroxymethyl) -. monosodium salt glycine; n- (hydroxymethyl) glycine, sodium salt; sodium hydroxymethylglycinate sodium hydroxymethylglycinate; sodium n- (hydroxymethyl) glycinate; sodium salt glycine, n- (hydroxymethyl) -; sodium salt n- (hydroxymethyl) glycine

Iodopropynyl Butylcarbamate- Thought to present possible risks to human reproduction and development, linked to the potential for reduced fertility or reduced chance for a healthy, full-term pregnancy.

Function: preservative

Synonym (s): 3-iodo-2-propynyl butylcarbamate; butyl-3-iodo-2-propynyl ester carbamic acid; butyl-3-iodo-2-propynylcarbamate; butylcarbamic acid, 3-iodo-2-propynyl ester; carbamic acid, butyl, 3iodo2propynyl ester; carbamic acid, butyl-3-iodo-2-propynyl ester; idopropynl butylcarbamate; iodopropyl butylcarbamate; iodopropynl butylcarbamate; ipbc

Phenoxyethanol- This is often made from natural sources and is used as an antibacterial and a preservative. This can be found in many organic skincare brands, vaccines, and bug repellants. It is also commonly known as ethylene glycol, phenyl ether or ethylene glycol monophenyl ether. This is often polluted with carcinogenic toxin 1,4-Dioxane.

According to research from the Journal of Industrial Hygiene and Toxicology, phenoxyethanol is linked to affecting the brain and nervous system in animal studies at modest doses. Back in the early 1990s, the Journal of the American College of Toxicology stated that phenoxyethanol performs like an endocrine disruptor. This could also cause damage to the bladder and cause acute pulmonary edema. Back in the early 1980s, animal studies suggested that it might even cause DNA mutations. Not much human research has been conducted on this matter. This chemical is a very well-known irritant to the skin and eyes. Many countries even have it classified as such or have restricted it. **"Phenoxyethanol is a preservative that's primarily used in cosmetics and medications. It also can depress the central nervous system and may cause vomiting and diarrhea."**

It might even be linked to cancer at high doses, and organ damage, developmental defects, and brain and nervous system side effects. It is a skin allergen and irritant. I have not found any studies that have researched the long-term effects of normal doses over years of use.

Function (s): Fragrance Ingredient; Preservative

Synonym (s): 2-hydroxyethyl phenyl ether; 2-phenoxy-ethanol; 2-phenoxyethanol; 2-phenoxyethyl alcohol; ethanol, 2-phenoxy-; ethanol, 2phenoxy; ethylene glycol monophenyl ether; phenoxytol; 1-hydroxy-2-phenoxyethane; 2-fenoxyethanol (czech); 2-phenoxyethanol

Section 21
Hidden Ingredients in Food!

Would you eat a bowl full of rat poison and call it cereal? I really don't think you would. What about giving your child deicing chemicals for breakfast? I think you would have to be insane in order to do that as well. I ask these questions because foods that claim to be "healthy" may contain harmful chemicals, many of which the average person has no idea what they really are.

Do you remember the 1997 Breyers ice cream commercial that asks a very simple question— "What's in your ice cream?" In the commercial, they have young children read the label of some brands of ice cream. The kids are unable to read what's in the ice cream, let alone understand what it means. The message that Breyers was trying to get across is that if you can't even pronounce what's in your food, how can you begin to understand what it's doing to your body?

To this day, manufacturers are finding ways to add new and unproven chemicals to their products. Many of the chemicals are only based on their own independent safety studies, without any approval by the FDA.

Red 3 (Erythrosine): Back in 1990, this dye was deemed by the FDA as a thyroid carcinogen by means of animal studies. It has been banned in cosmetics and topically applied drugs. Even though it was banned in those products, it can still be found in many food products such as sausage casings, oral medication, maraschino cherries, baked goods, candies, and many more.

Red 40 (Allura Red): This is one of the most commonly used dyes. It has increased the rates of immune system tumors in lab mice, it is also linked to hypersensitivity in some of the consumers. This dye is commonly found in beverages, bakery goods, dessert powders (jelly, puddings), candies, cereals, foods, drugs, and cosmetics.

Azorubine, Carmoisine (also known as E122): This is a synthetic coal tar dye. It has been linked to side effects in people who are asthmatics and people allergic to aspirin. This dye has been linked to cause hyperactivity, also called "ADHD," in children. This dye is commonly found in soda, marzipan products, preserves, jelly/jams and preserves, Swiss roll, sweets, brown sauce, flavoured yogurts and packet soups, breadcrumbs and cheesecake mixes.

Blue 1 (Brilliant Blue, FCF): studies have liked blue dye 1 to low blood pressure, hives, and allergic reactions. suggested that the dye Blue 1 and other FD&C food colors is noted to absorbed more and faster, blue 1 has been noted to cross the blood-brain barrier. Blue 1 was once used on patients that was on enteral feeding because they were unable to eat normally. This dye can be found in soft drinks, jello, ice cream, drink powders (cool), candy, bakery products, cereals, feta cheese, dairy products and pudding. Also used in toothpaste, mouthwash, deodorants, cosmetics and pet foods.

Blue 2 (Indigo Carmine): On September 2007, a study reported by D. McCann in the journal *"The Lancet"* had linked blue 2 to ADHD in children. A group of studies reviewed by the *Center for Science in the Public Interest,* Blue 2 had a statistically significant increases in brain cancers and other abnormal cell development. It is to be noted that CSPI (*Center for Science in the Public Interest*) asserts that Blue No. 2 is not safe for human consumption.

Caramel: There are four different kinds of caramel coloring. They lump them all in the labeling and don't state which of the four you're really ingesting. Caramel color is by no means made with caramel candy. It is a product made by means of heating sugars with ammonia and sulfites. Some caramel coloring contain a potentially carcinogenic chemical called 4-methylimidazole (4-MeI). In 2011, the federal government conducted a study on 4-Mel, this the *International Agency for Research on Cancer* determined that the chemical is **"possibly**

carcinogenic to humans." Currently there are no limitations in place for 4-Mel and in foods or beverages. Not all sodas have the same amount of 4-Mel, here is what one study found: **"For example, for diet colas, certain samples had higher or more variable levels of the compound, while other samples had very low concentrations,"** says Tyler Smith, lead author of the study and a program officer with the *Center for a Livable Future*.

"Until now, how much of these neurotoxic chemicals are used in specific foods was a well-kept secret," said CSPI executive director Michael F. Jacobson. **"I suspect that food manufacturers themselves don't even know. But now it is clear that many children are consuming far more dyes than the amounts shown to cause behavioral problems in some children."**

Citrus Red 2: This dye is toxic to rodents even at low levels and caused tumors. Studies show this additive causes cancer.

Green 3 (Fast Green): Caused significant increases in bladder and testes tumors. It's commonly found in drugs, personal care products, cosmetic products, candies, beverages, ice cream, sorbet, and ingested drugs.

FD&C Yellow 5 (Tartrazine, E102, FD&C Yellow 5, C.I. 19140, or just plain Yellow 5): As of April 1st, 2013, the Federal Regulations state that the label used for yellow 5 must include a warning. This color, as many of the others, is a known carcinogen. The warning label must state that the color additive may cause allergic reactions, one of which is asthma. Yellow 5 shouldn't be consumed by individuals who have an aspirin hypersensitivity. The connection between sensitivity to yellow 5 and aspirin has been stated in numerous studies.

Yellow 5 can cause hyperactivity in some children. The Food Standards Agency (FSA), which is comparable to the FDA, warned back in 2008 that some food coloring such as yellow 5 can cause behavioral changes in children. This is what we would call ADHD. The FDA recommends avoiding or restricting the amounts of yellow 5 that are consumed. They also state that if a child is showing signs of hyperactive behavior to simply remove yellow dyes from the child's diet and see if there are any changes. Keep in mind, it could take time for the dye to be completely removed from their system.

Yellow 5 has been linked to thyroid tumors. This could lead to weight problem, heart palpitations, high blood pressure, diarrhea, development of a goiter (a neck enlargement), and painful or heavy periods. Consult your doctor if you suspect you might have a thyroid problem.

FD&C Yellow 6 (also known as Sunset Yellow, FCF, and Orange Yellow S). May cause allergies, hyperactivity, and chromosomal damage. This coloring is used in many food products such as cakes, candy, pork sausage, and gelatin desserts. Yellow number 6 can cause hives in some people. Some evidence suggests that it could increase the chance of bruising. The research shows that this color causes indigestion, vomiting, runny nasal congestion, and may also contribute to kidney and adrenal gland tumors.

Annatto 160b (Arnatto, Roucou, Achiote, Bixin and Norbixin): This dye is red/yellow in color it is commonly used as a food dye, textiles, meat preservative, and beauty products. Annatto has been reported to affect your skin, gastrointestinal, respiratory, and central nervous system damage it is commonly mixed with MSG.

Professor Floch is *Clinical Professor of Medicine at Yale University School of Medicine*. His review concludes **"It is clear that annatto is common in our foods, it is clear that it is not known as a significant producer of allergic responses, and it is unknown to most of our practitioners dealing with allergies and the irritable bowel syndrome. Therefore, it behooves us to begin studies in investigating the role of dyes such as annatto in the production of the symptoms of the irritable bowel**

syndrome."- Floch MH. Annatto, diet, and the irritable bowel syndrome. J Clin Gastroenterol. 2009;43(10):905-6.

Annatto is found in American cheese, cheddar cheese, and velveeta cheese, some crackers, cereals, some salad dressing, some ice creams, gourmet mustards, bouillon cubes, sugar-free Jello, crystal light mixes, some butters, microwave and theater popcorn.

If you're looking for dependable and safe ways to dye your food and drink, the alternatives to the synthetic dyes are safe and easy to produce and buy. The natural forms of dyes are plant-based fruits and vegetables, so you know they are healthy and organic and non-GMO. These kinds of dyes have been practiced for thousands of years and won't change the taste or aroma of your food or drink.

The most important thing is they are not linked to causing any adverse outcome such as allergic reactions. Many of the ancient natural dyes for coloring are not merely beneficial to your health, but are linked to the restoration of cells and innate healing. In this section, we'll have a look at some of the companies that make natural and safe food coloring.

Wild: At Wild, they offer safe, natural food and beverage colors that are made with organic materials. They are an active member of the International Association of Color Manufacturers (IACM), which regulates purity and dependability. You can find Wild at wildcolours.co.uk.

Seelect Tea: Seelect's are USDA organic certified food colors that are made with various fruits and vegetables such as cabbage, hibiscus flowers, gardenia flowers, annatto seeds, turmeric root, beets, carrots, and cabbage. You can find Seelect tea at seelecttea.com they ship international.

Nature's Flavors: With a very wide range of colors made from flowers and vegetables that contain antioxidants, bio-flavonoids, and polyphenols, Nature's Flavors are organic and are one of the first companies to produce all-natural organic flavors and colors. You can find Nature's Flavors at naturesflavors.com.

Here is what to look for on the packaging to make sure your dyes are safe:

Natural YELLOW—made from turmeric powder, yellow carrots, lemon zest, saffron flowers, bee pollen.

Natural ORANGE—made from orange carrots, orange zest, carrot juice, pumpkin, papaya, cumin powder.

Natural RED—made from red beets, chili powder, strawberries, pomegranate.

Natural PINK—made from beetroot, pomegranate powder, raspberries.

Natural PURPLE—made from red grapes, organic red wine, red cabbage, purple carrots, purple potatoes, Acai berry powder.

Natural GREEN—made from spinach juice, wheatgrass, kale, swiss chard leaves, lime zest.

Natural BLUE—made from blueberries, blackberries.

Make Your Own Food Coloring

There are many ways to make your own food coloring; these methods are safe and healthy.

Yellow

Bring 2 cups of water to a boil.
Let the water cool for one minute.
Add a small amount of turmeric to the water.
Keep adding small amounts of turmeric until you reach the desired color.
After cooled, keep in a glass container.

Red

Place in a medium-sized pan several unpeeled beets and cover them with water.

Cook beets for about 30-35 minutes or until a fork can come out with ease.

Remove the pan from the heat and let it cool.

Peel the beets and chop into small pieces.

Place beets back into water.

Leave the beets and water 5-10 hours or overnight.

Filter the liquid through a piece of cheesecloth or paper filter into a glass jar.

Add 2 teaspoons of organic white vinegar to the water.

Shake well and store.

Green

Put 2 cups of fresh spinach leaves in a pot.

Cover the leaves with water.

Boil the leaves for one minute.

Let cook for ten minutes.

Allow the water to cool.

Filter the colored water through a cheesecloth or paper filter into a glass jar.

Store with a tight-fitting lid.

These colors can be mixed to get the color you want and are great for Easter. They are fun to make and take very little effort.

Section 22
Real Dangers that are Lurking in your Home

This list contains some of the most common ingredients in food. Many of the chemicals are known carcinogens, toxins, hormone disruptors, poisons, and contaminants.

Acetone peroxide is very explosive, this chemical is made from hydrogen peroxide, acetone and hydrochloric acid. Used in the food industry as an oxidizing agent, it is commonly used in flour for making it white, normal flour is a yellowish color, over time the flour will turn white. This aging process takes to long for the mass production of flour so acetone peroxide is added. Because Acetone peroxide is made by mixing acetone with hydrogen peroxide this can have the same health effect as acetone this just take more time to have an effect on your health.

Inhalation of moderate to high amounts, even for a short time, results in the entry of acetone into the bloodstream, where it is channeled to all other organs. It's a nose, pharynx, lung, and eye irritant. It can cause headaches, confusion, increased heartbeat rate, effects on blood, nausea, vomiting, unconsciousness, and coma. It shortens the menstrual cycle in women. Effects of long-term exposure include kidney, liver, and nerve damage, increased birth defects, metabolic changes, and coma.

Used to whiten, mature, and strengthen flour.

Alkyl-phenol Ethoxylades also known as (APEs) are toxic, however, when they are broken down from use they are called alkyphenols (APs) are considered to be 10 times more toxic. They are a known hormone disruptor because they mimic estrogen, the bad thing is they are not biodegradable can be found in drinking water.

"APEs do not biodegrade easily after they are washed down the drain. As a result, nonylphenol has been found in

water and sediment downstream from sewage treatment plants, paper pulp mills, and industrial facilities. Some studies have found altered reproduction, feminization, hermaphroditism, and lower survival rates in salmon and other fish living in nonylphenol-contaminated water. These effects have been found in wildlife even at low doses."

The scary thing is that APEs are also bioaccumulative, this means that once they get in your body they accumulate in its tissues over time. APEs have been found in human blood and breast milk. Watch out for products containing chemicals whose names end in **phenol ethoxylate.** May reduce sperm count. Found in shampoo, bubble bath, detergents, cleaning products, pesticides, lubricants, hair dyes and other hair care products, and even spermicides.

Benzene is used in many detergents, drugs, pesticides, and adhesives. Benzene is linked to causing headaches, rapid heart rate, tumors, confusion, unconsciousness and even death. Benzene is linked to Hodgkin's and lymphomas as a result of inhalation. All of these side effects are when exposed to high levels. Not many long-term studies have been conducted on inhaling small doses over time.

Bronopol is a very well-known human immune system toxicant. It has been connected to lung and skin damage. Even at low doses, bronopol has been shown in animal studies to have an effect on the brain and nervous system, gastrointestinal tract, and has extensive universal effects. This ingredient is restricted for use in cosmetics in Canada and listed by the European Union as a toxin, but it is still in use in the USA. Bronopol is commonly used in cosmetics and baby wipes.

Function (s): Preservative

Synonym(s): 2-bromo-2-nitropropane-1,3-diol, 1,3-propanediol, 2-bromo-2-nitro-; 1,3propanediol, 2bromo2nitro; 2-bromo-2-nitro- 1,3-propanediol; bronopol; 2-bromo-2-nitro-1,3-propanediol;

2-bromo-2-nitropropan-1,3-diol;   2-bronopol;
beta-bromo-beta-nitrotrimethyleneglycol;   bronidiol;
bronocot; bronopol

Butylparaben is commonly known as methylparaben,
ethylparaben, propylparaben, and butylparaben. Some of the
lesser-known, but still-used parabens include isobutylparaben,
isopropylparaben, and benzylparaben. Studies have shown
parabens to mimic estrogen. The studies have also shown a link in
the development of several forms of breast cancer. Parabens mimic
estrogen, it also lowers the development of testosterone and the
function of the male reproductive system. Other side effects
include itching, burning, and blistering of skin.

Found in body care products.

**C. Cohnii Oil (Crypthecodinium Cohnii Oil) and M.
ALPINA OIL.** Created by Martek's, DHA and ARA products are
synthetic efforts to make a human breast milk used in the U.S.
market since 2002, and inorganic products since 2006. They are
extracted from dinoflagellate microalga red algae and fungi that's
not in any way part of the human diet, and have never been
approved by the USDA. The company "Martek" synthetic
omega-3 is separated using a toxic petrochemical solvent called
hexane, which is a chemical made from crude oil. DHA causes
severe gas, diarrhea, vomiting, gastric reflux, constipation, bowel
obstruction, agitation, fussiness, crying, and severe distress.

C. Cohnii Oil is commonly found in baby formula and infant
cereals.

Carrageenan is toxic and causes inflammation in the
digestive system, and is listed as a potential human carcinogen by
the World Health Organization. This may lead to harmful effects
on human health, including inflammation, lesions, and cancer in
the colon.

Synonym(s): Danish agar (from Furcellaria fastigiata),
Eucheuman (from Eucheuma spp.), Furcellaran agar (from
Furcellaria fastigiata), Hypnean (from Hypnea spp.),

Iridophycan (from Iridaea spp.), Irish moss gelose (from Chondrus spp.)

Commonly found in: toothpaste, gummy products, dairy products/plant, milks, beer, shoe polish, shaving cream.

Hydroquinone is a severely toxic chemical that has been banned in the United Kingdom, but is still used in the U.S. Research has linked this chemical to skin cancer. It is commonly found in skin bleaching and lightening products.

Other side effects of hydroquinone:

• Skin has an itching sensation that won't go away

• The feeling of the skin burning remains present at all times

• The treated areas can begin to swell

• Large skin rashes develop

• Other treated areas turn crusty and hard

• The skin can become a discolored blue or black, also known as ochronosis

• Numerous topical hydroquinone dangers lead to severe allergic reactions and can occur on the skin or inside of the body

Magnesium Stearate/Stearic Acid is formed by adding a magnesium ion to citric acid. This additive is found in 95% of supplements today. Magnesium stearate is absolutely useless to your body; a metal derivative that your body has absolutely no conceivable use for. Magnesium stearate or stearic acid is basically a toxin—a combination of hydrogenated oils that gets into your body and starts killing cells almost immediately. Magnesium stearate may be loaded with pesticides that are employed in the cottonseed oil that has been hydrogenated. This can be found in pharmaceuticals, vitamins, foods, talcum powder, ammunition, and as a drying agent in paints.

Olestra/Olean, more than 15,000 consumers filed complaints within a short time shortly after it was released to the public. The plaintiff, a 30-year-old woman from Braintree, Mass., she had experienced diarrhea, stomach cramps, and other symptoms soon after eating Ruffles Light potato chips.

"It's bad enough that Frito-Lay still uses this discredited and dangerous chemical, one of the most infamous food additives in history," said CSPI litigation director Steve Gardner. **"But by quietly changing the name of this product line and purposefully deemphasizing the presence of olestra, Frito-Lay is really tricking consumers. And that deception is putting Americans at risk of some pretty unsettling side effects. CSPI has taken this action only after many months of informal efforts to convince the company to take corrective action voluntarily."**

This chemical doesn't allow the body to absorb vital nutrients causing vitamin deficiency while flushing them out of the body. One of the nutrients called "carotenoids" that's flushed away helps fight off diseases such as lung cancer, prostate cancer, heart disease, and macular degeneration Laxative-like effects Olestra replaces fat in "fat-free" foods. Chips in Frito-Lay's "Light" line include Doritos Light, Lay's Light original and barbecue, Ruffles Light original and cheddar and sour cream, and Tostitos Light.

Perchlorate, study published in 2005 by the National Academy of Sciences panel determined that this chemical affects the thyroid's ability to take in iodine. Women especially are at risk of hypothyroidism due to a common exposure to the toxin. Perchlorate is a byproduct of rocket fuel and fertilizers. This chemical has been found in the nation's drinking water supply, as well as fruits, dairy products, vegetables, and grains irrigated by perchlorate-contaminated water. This problem is not just in the drinking water. A study that was conducted in 2005 by Texas Tech University showed that they had collected 36 samples of breast milk from women in 18 states. All of them contained perchlorate, which means that these women were passing the chemical to their child. The study was also conducted on 47 cows' milk samples. Out of those, only one did not contain perchlorate.

The big problem is that there is no real report on perchlorate in the water, and many states don't even report the levels in their own state to the EPA or any other federal body. This is clear in a report by the GAO: **"State and other federal agencies don't always report perchlorate detections to EPA, however, because EPA, other federal agencies, and the states don't have a standardized approach to reporting perchlorate data nationwide... Further, EPA doesn't track the status of cleanup at sites where perchlorate has been found. Without a formal system to track and monitor perchlorate findings and cleanup activities, EPA and the states don't have the most current and complete accounting of perchlorate as an emerging contaminant of concern, including the extent of perchlorate found and the extent or effectiveness of cleanup projects."**

If you're looking to remove the perchlorate from your home's drinking water, there is good news. Perchlorate can be removed from drinking water through reverse osmosis. There are home treatment units that are certified to remove perchlorate from drinking water. Not even organic foods are safe from the chemical perchlorate. So the question is, what can I do about this?

There are three main things a person can do to protect themselves and their families:

1. Iodine—this chemical is very well known to block and reduce the levels of perchlorate in the body. Take iodine everyday, especially if you're breastfeeding or pregnant. Iodine is good for the infant's brain and organ development, so it is safe.

2. Try as much as you can to drink water that's filtered and clean. Most reverse osmosis water will reduce this chemical to safe levels.

3. Call your water supply company and ask for the levels in your drinking water to see what they are.

PFOA or C8 and PFOS (also known as PFCs) these two chemicals are well known to be very toxic, so much so that they are to be phased out by 2015. Even as the EPA, and DuPont the makers of this chemicals have agreed to remove them from consumer products, you are still being exposed in your drinking water. DuPont agreed to slowly remove products containing these two chemicals through 2015, even then, many people are going use products that may have these chemicals in them. According to a study published in *Environmental Health Perspectives* back in 2010, high levels of PFOA within the body could lead to thyroid disease. They have found that women could have a harder time conceiving children than women with lower concentrations of the chemicals.

2009 study published in *Human Reproduction*. The research showed that the likelihood of infertility went up by 60% to 154%, if the levels of PFOA were high enough. You can lower your exposure to PFOA and PFOS by avoiding things such as nonstick pans, food packages ie. microwavable popcorn and chip bags, fast-food wrappers many fast food boxes and paper wrappers are coded with one of this chemicals, pizza boxes, stain-repellent rugs, upholstery, and even cosmetics. Women are not the only ones who can have their health affected by PFOA's. Men are affected as well, men who have tested positive had lower sperm counts, according to a 2009 study in *Environmental Health Perspectives*.

Manufacturers of PFOA expected to finish producing the harmful chemical by 2015. PFOA is linked to significantly increasing women's risk of developing arthritis, according to a survey led by researchers from Yale University, Harvard Medical School, Brigham and Women's Hospital, and published in the journal *Environmental Health Perspectives*.

"We found that PFOA and PFOS exposures are associated with higher prevalence of osteoarthritis,

171

particularly in women, a group that's disproportionately impacted by this chronic disease," researcher Sarah Uhl said.

Studies done in the past have also linked the chemicals to premature menopause. **"Once they get into the environment, they just don't go away,"** Sarah Uhl said. **"In people, they last years. So even if we were to reduce the use of these chemicals right away, they're still going to be around and in our bodies for a long time."**

Sodium Nitrite back in the 1970s, the USDA attempted to ban the use of sodium nitrite. The food companies voted for the use of sodium nitrite, as they didn't have any other means to preserve meat and protect it from foodborne bacteria. The concern about sodium nitrite is that the N-nitrosamines are made when sodium nitrite reacts with amino acids, such as the ones in your stomach. Not only are they linked to causing tumors, but they get into the bloodstream and damage organs like the liver and pancreas. Sodium nitrite may also trigger migraines. High consumption of sodium nitrite could lead to certain types of cancers, like gastric cancer, esophageal cancer, colorectal cancer, and colon cancer.

Sodium Nitrite are using in almost all forms of meat products, even precooked meats can contain sodium nitrite, look for organic meat product or products that have "No nitrites."

Triclosan, studies on the use of triclosan on animals are making scientists concerned. It could increase the risk of infertility, early puberty, and other hormone-related problems in humans.

"To me, it looks like the risks outweigh any benefit associated with these products right now," said Allison Aiello, professor at the University of Michigan's School of Public Health. **"At this point, it's just looking like a superfluous chemical."**

Triclosan is known to cause abnormalities with the endocrine system, particularly with thyroid hormone signaling; weakening of the immune system; birth defects; uncontrolled cell growth; unhealthy weight loss. It can be found in antibacterial soaps and body washes, toothpastes, and some cosmetics.

I think that every shopper should start reading the labels and understanding what is in the product you're buying. You will find that searching for organic ingredients will benefit your health and be well worth the effort.

Section 23
Sunscreen Danger!

Every year, millions of people lather on sunscreen, thinking it could protect them from skin damage and cancer. Every person is taught that the sun is the main reason for skin cancer. I started thinking a long time ago—if sunblock is to prevent skin cancer, why is the sun still to blame if we are adequately protected? The way sunblock is meant to work is by blocking the damaging UV and UVB rays. For many years now, the media would have you believe that you're to use sunblock every time you're going to be exposed to the sun for a long period of time.

They claim that children are most important and should have sunscreen applied before going out into the sun. The manufacturers claim that sunscreen can protect your skin from wrinkles and sunspots. A study done over fifteen year duration found the claims to be untrue. The chemicals within sunscreen could actually cause cancer. The same research has shown that sunlight may in fact reduce cancer rates and improve your overall health.

There is an overwhelming number of concerns about exposure to radiation from the sun and the risk of skin cancer. This has resulted in the masses using sunscreen, many of the name-brand sunblocks contain chemicals linked to being estrogen mimickers and also lipophilic (loving fat and binding to it).

Not only have these chemicals been found in cosmetic and sunscreens, but there has been definitive proof that the same chemicals have been found in human urine and breast milk after use. 5 out of 6 sunscreens tested show substantial estrogen levels. These were measured against the proliferation rates in human breast cancer cells.

The chemicals that increase estrogen are: 3-(4-methylbenzylidene) -camphor (4-MBC),

174

Octyl-Methoxycinnamate (OMC), Octyl-dimethyl-PABA (OD-PABA), bexophenome-3 (Bp-3), and homosalate (HMS).

A study conducted a few years ago have noted that using Octyl-Methoxycinnamate on the skin of lab rats increased the absorption of the endocrine-disrupting herbicide 2,4-D. There are many organic sunblock that have no damaging chemicals and use natural means to keep you and your family safe. Two good organic brands to look for are Badger Balm and Raw Elements. Both of them are organic and safe, and they have a wide range of products to fit the whole family.

The following sun-filtering agents has been implicated with possessing estrogenic effects:

• Benzophenone-1, 2, 3 and 4 (also known as Oxybenzone).

According to a study conducted in 2013 by CNN, about 25% of the sunscreen that's sold in today's market can actually protect you from the damaging sun rays without using harmful and damaging chemicals. The other 75% contain very damaging ingredients that are toxic. Oxybenzone was approved by the FDA back in 1978 and is in half of the sun protection products on the market. This includes the lip balm variety as well. A study by the Environmental Working Group suggest that we actually absorb oxybenzone through our skin, which could cause hormone imbalances, cell damage, and maybe even skin cancer. That's right—a product that was developed to protect your skin can cause the same thing it was designed to protect.

• Padimate-O and Parsol 1789

These chemicals damage DNA when it is illuminated by sunlight. Dr. Knowland, a biochemist at Oxford University, studied the effect of these chemicals and advises, **"DNA damage inflicted by an excited sunscreen is much less capable of being repaired by naturally occurring repair**

mechanisms than the DNA damage inflicted by UV alone.”

• Homosalate is a weak hormone disruptor

• Octyl Methoxycinnamate (Octinoxate), causes developmental and reproductive toxicity through enhanced skin absorption

• 4-Methylbenzylidene camphor UV-B 4-MBC, Eusolex 6300, are linked to thyroid toxicity, hormone disruption.

• Triethanolamine is found to be a skin sensitizer, eye irritant, respiratory irritant, asthma trigger, human immune system toxin, slight human carcinogen, and a slight organ toxin. Triethanolamine effects are enhanced if the products are left on the skin, such as moisturizers.

Some of the safe ingredients to look out for are:

Avobenzone is commonly used and effective against sun filtering.

Physical blocking agents zinc oxide and titanium dioxide have not been linked to any adverse health effects or any estrogenic side effects.

Mexoryl SX, Mexoryl XL, and Tinosorb appear to be safe.

The FDA's 2007 draft on sunscreen safety regulations clearly stated: **“FDA is not aware of data demonstrating that sunscreen use alone helps prevent skin cancer.”** The *International Agency for Research on Cancer* (IARC) writes that **“sunscreens shouldn't be the first choice for skin cancer prevention and shouldn't be used as the sole agent for protection against the sun.”**

In conclusion, sunblock as protection from the sun's radiation, causes the same thing that it's designed to protect against, but is linked to many more effects than the sun alone. I wouldn't use sunblock unless you really do your research on each ingredient and understand the health risks. Organic sunscreen are always the best option, just be sure to look at the ingredients and understand what they are before you use the sunscreen.

Section 24

Is Your Soda Killing You?

A lot of people drink soda on a daily basis, over the past few months, videos have been swirling all over the Internet about what happens if you cook your soda, urns into the thick black tar for those who may not have seen the videos. Do you know what soft drinks were made for originally? The origins for soda pop are actually found in pharmacies—it started out as a drug or pharmaceutical substance. John Pemberton, American pharmacist, is best known for being the inventor of Coca-Cola. He patented his soda pop as medicine. Coca-Cola was known to relieve indigestion and dyspepsia (discomfort in the upper abdomen or chest that may be described as gas, a feeling of fullness, gnawing, or burning).

7-Up was originally used on mental patients and was commonly given to people in the mental hospital. It was named 7-Up because the drink only had seven ingredients, one of which was lithium—a known chemical element that's still in use to this day for depression. Even though 7-Up has not contained lithium since 1950, over the years, we have become accustomed to the taste. Nevertheless, it is still very dangerous to our health.

All health experts are in an agreement that we all should be eating between 1,750 and 2,100 calories per day. This is to ensure that we get a balanced diet, a grown man should be eating about 150 calories from sugar, and a woman about 100. A single can of soda has about 130 calories of sugar, and there are some that have far more. Just a 16 oz. can of soda is twice the sugar we should be ingesting. As we are on the topic of sugar, what kinds of sugar or sweeteners are in soda? Soda pop is not known for being the healthiest drink. It harbors some of the most harmful synthetic sweeteners we have ever known.

Aspartame (L-Aspartyl-L-Phenylalanine, Methyl Ester) is 200 times sweeter than sugar. This sweetener is made by combining two amino acids and aspartic acid. It can contain up to 10% methyl alcohol. This additive is known to be poisonous and is used as a solvent and antifreeze. It's an addictive carcinogenic neurotoxin and harms brain health and function. It's linked to harming the brain development in unborn babies.

If aspartame is heated even by leaving it in a car on a hot day and reaches above 86 degrees Fahrenheit, it transforms into a deadly neurotoxin known as formaldehyde. If you think about your body, its internal temperature is 98.6 degrees Fahrenheit, which means that aspartame will automatically convert into formaldehyde once inside the human body.

Saccharin (benzosulfimide, Sweet' N Low, Sweet Twin, Necta Sweet) Even though saccharin doesn't change in the body or with heat, it is just as harmful. Saccharine is linked to causing bladder cancer in animal testing, which suggests that it will also cause bladder cancer in humans.

Sugar—sugar is used in soda, but how do we know if is real cane sugar of the man-made white table sugar. We'll discuss in more detail why that makes a big difference in the sugar section of this book.

Dye—in the dye section of this book we have talked about the health dangers of Caramel coloring, there are numerous sodas that contain one or more food coloring in them. Some soft drinks have very high amounts of a drug called caffeine, which is part of the family known as methyl-xanthines. Caffeine reduces you in B vitamins, more specifically B1 (thiamin). This reduction of the essential vitamin can lead to symptoms including fatigue, nervousness, general aches and pains, and headaches. Overuse of caffeine causes the food in your digestive tract to ferment, which could lead to your body becoming poisoned by the rotting undigested food. This can lead to upset stomach and diarrhea with a foul smell. Caffeine also has diuretic properties (forces the production of urine). This puts unneeded stress on the kidneys and could lead to dehydration. Studies conducted on the long-term use of caffeine state that it restricts the DNA replication. That could have links to birth defects caused by vital nutrients not being absorbed by the mother or the unborn child.

Soda is one of the main reasons for obesity, but being overweight is not the only danger of soda pop—it is linked to metabolic syndrome. According to the New York Times: **"A major, if not the major, risk factor for heart disease and diabetes. The Centers for Disease Control and Prevention now estimate that some 75 million Americans have metabolic syndrome. For those who have heart attacks, metabolic syndrome will very likely be the reason. The first symptom doctors are told to look for in diagnosing metabolic syndrome is an expanding waistline. This means that if you're overweight, there's a good chance you have metabolic syndrome, and this is why you're more likely to have a heart attack or become diabetic (or both) than someone who's not."**

We have talked about how cancer needs an environment with a low level of oxygen, and that high amounts of sugar leads to diabetes and obesity. Mark Bittman of the New York Times reported: **"After accounting for many other factors, the researchers found that increased sugar in a population's food supply was linked to higher diabetes rates independent of rates of obesity."** This in turn depletes the body of oxygen and making the perfect home for cancer to grow. In the article, Bittman discussed the overconsumption of processed sugar. **"One of the diseases that increases in incidence with obesity, diabetes, and metabolic syndrome is cancer... The connection between obesity, diabetes, and cancer was first reported in 2004 in large-population studies by researchers from the World Health Organization's International Agency for Research on Cancer. It is not controversial. What it means is that people are more likely to get cancer if they're obese or diabetic than if you're not, and more likely to get cancer if you have metabolic syndrome than if you don't."**

The way that sugar depletes the oxygen in your body is by the fermentation of sugar, the results of such a process is a release of gases. These gases replace oxygen in normal body cells from the fermentation of sugar.

Is diet soda better?

As people become more and more health conscious, there seems to be confusion about what is healthy. The media are very good about not informing you with the truth and adding to that confusion. The soda manufacture adds labels like "sugar free," but neglect to tell you that they have replaced the sugar with something much worse. Understand what the replacement for the sugar is and read the label.

Section 25
Types of Sweeteners and What They Are

Nutritionists agree that you should eat all forms of sugar in moderation, but almost in the same breath, they claim that you can eat all the fruit and vegetables your heart desires. Don't they hear themselves? If we eat all the fruit and vegetables we want, but limit the sugar, isn't that a little contradictory? All fruits and vegetables have sugar!

It's human nature to want sweets. Even back in the hunter and gatherer days, we consumed some form of sugar. The danger is that many of the so-called sugars or sweeteners we have today are not real sugar—they're man made poisons passed off as sweeteners. In this section, we'll be talking about the many forms of sugars and sweeteners, including the chemicals that are contained within them and what they're doing to the human body. Many people think that they're created the same and they're relatively healthy, but that's not the case.

It's easy to categorize sweeteners into six groups, as follows:

Sugar alcohols

Caloric sweeteners

Natural sugars

Natural zero calorie sweeteners

Modified sugars

Artificial sweeteners

Sugar Alcohols

These forms of sugars can be found naturally, just not in large quantity. They have fewer calories per gram than sugar, and most importantly, are not known to cause tooth decay. Many sugars alcohol come with some unpleasant side effects if consumed to excess. They are known to cause bloating and diarrhea, this form of sugar acts as a laxative. We are going to look at all of them and see if they are safe or not.

Sorbitol (also known as glucitol) used to be a widely accepted alternative for diabetics and the health conscious. Research has suggested that sorbitol is deemed unfit for human consumption. Sorbitol depletes the body of essential nutrients like vitamins, minerals, and amino acids. Sorbitol has the ability to draw water into cells, leading to swelling in the cells. This reaction can eventually result in serious health problems such as diabetic neuropathy. Diabetics are more likely to be harmed.

Here is what you should be aware of:

Vision problems

Kidney problems

Blood vessel damage

Xylitol (also commonly known by as birch sugar, E967, meso-xylitol, méso-xylitol, sucre de bouleau, xilitol, xylit, xylite, xylo-pentane-1,2,3,4,5-pentol). Xylitol can cause diarrhea when consumed in large amounts, according to Yale-New Haven Hospital. This is most likely to happen when consumed at more than 40 g per day, according to Epic Dental of Provo, Utah. **"RXList.com reports that it's safe for adults to consume up to 50 g of xylitol each day, but that people need to avoid higher doses. There is concern that taking xylitol in extremely high doses for more than three years may cause tumors. Kids need to be restricted to 20 g per day."** In small doses, it seems to be okay.

Mannitol, People are unaware that mannitol is used as a drug in preventing or treating body water retention and certain kidney conditions, reducing swelling of the brain, or reducing pressure in the eye.

Less serious side effects may include:

Nausea, vomiting

Runny nose

Dizziness

Chills

Mild skin rash

Erythritol (also known as crystalline fructose), As with all sugar alcohols, this lead to abdominal pain, bloating, and diarrhea, it is not easily absorbed and digest by the body. Individuals with intestinal disorders should avoid using erythritol. People who have IBS or cancer should avoid erythritol, as it can make the symptoms worse and promote the growth of cancer in individuals who already have it.

Erythritol is a fructose based sweetener that is known to significantly raise your triglycerides, this dramatically increases the risk of heart disease. Studies conducted on erythritol have found arsenic, lead, chloride, and heavy metals within the sweetener.

Here is the known side effects of Erythritol:

Diarrhea

Headache

Stomachache

Isomalt (also known as Palatinitol; Isomaltitol; Palatinit) was discovered in the 1960s, but it didn't become popular until recently. It contains 2.1 calories per gram. Over Consumption can lead to painful upset stomach, bloating, and gas. Some people reported this effect even without consuming isomalt in large quantities. This is because the body can't break this sweetener down, it's recommended to only ingest about 50 g for adults and 25 g for children. Studies suggest that the use of isomalt over time will lower your tolerance, making it have more effect on your body.

Lactitol (also known as4-O-(B-galactopyranosyl)-d-glucitol; 4-o-beta-d-galactopyranosyl-d-glucitol) Lactitol has been used in the pharmaceutical industry as a laxative and can lead to upset stomach, bloating, and gas. Studies show that it can be passed from mother to child through breast milk with mild absorption.

Glycerol (ester of wood rosin) has a mild laxative effect and claims to help with dehydration. Glycerol has been found to be helpful in the treatment of cancer as glycerol becomes a usable cancer therapy that has particular value and complements antiangiogenic measures.

Caloric Sweeteners

Caloric sweeteners have been used for thousands of years. Over the past decade or so, they have been given a very bad reputation others are classified as healthy without deserving that title. They are not to be confused with the refined sugars—they are all in their natural form, and many have health benefits. As with anything, having too much can affect your health.

Organic Cane Sugar

Unrefined raw sugar is made from the juice from the sugarcane plant and has trace minerals and nutrients. Cane Sugar is not to be confused with white table sugar or refined sugar, unrefined organic cane sugar does not have empty calories, as many people think. In fact, it is quite beneficial for your health, it has 17 amino acids, 11 minerals, 6 vitamins, and includes antioxidants. Unrefined sugar cane itself has calcium, chromium, cobalt, copper, magnesium, manganese, phosphorus, potassium, and zinc. Sugar cane is rich in vitamins A, C, B1, B2, B3, B5, and B6, phytonutrients, chlorophyll, antioxidants, proteins, and fiber.

Organic Brown Sugar

Brown Sugar is made with organic cane sugar and molasses. Organic Brown sugar has fewer calories because of the water content, brown sugar is packed full minerals, it's known to help with blood circulation, enhance digestion and increase the production of blood cells. Organic brown sugar is great to help women with menstrual cramps, brown sugar has even been used to aid the recovery process of childbirth and uterine contractions.

Organic Honey

Honey has been used for thousands of years, not only as a food additive, but also for its healing properties and allergy relief. Research published by the National Institutes of Health, honey can be applied to infected wounds, which reduces redness, swelling, increases healing, and reduce the effects bacterial infection. It has been proven in treating conjunctivitis (an infection of the eye). The New York Times stated that honey is great in treating small, non-serious burns. It can be more effective than certain antibiotic ointments.

Here are a few little-known facts about honey:

• Honey has been used for relieving morning sickness in women for decades.

• It's known to help with sore throats, laryngitis, and pharyngitis.

• For people who have chronic bladder infections, try using honey and some cinnamon. Also, this remedy is great for arthritis, upset stomach, and bad breath.

• If you have migraines, honey and apple cider can help.

Honey has a natural form of botulinum endospores, because of this, you shouldn't give it to children under one year of age, as it can affect their immune system. There is some risk involved with raw honey—in its natural form, there might be some bee parts in the honey, like wings. There also may be pollen from plants that you may be allergic to. Even though raw honey is healthy, some people might be squeamish, knowing that they might be eating the critters that made it.

If you are scared of the small bee parts you are able to strain the raw honey by using a very fine wire mesh, I would not recommend using cheese cloth I have used this before and the cloth will absorb a good amount of the honey.

Organic Maple Syrup

Some people believe that all maple syrup is naturally organic and adding an organic USDA label is unneeded, but in fact, there is a big difference. Organic maple syrup is 100% free of any pesticides and chemicals, and it also means there is no formaldehyde. Yes, it is common practice for maple farmers to use formaldehyde. After they tap the tree, the hole will heal in about 4-6 weeks, so the farmers use formaldehyde so the hole doesn't heal and the sap still flows. This may damage the health of the tree and lower the health benefit of the syrup. Not only that, you're eating formaldehyde! Maple syrup is packed full of nutrients, including water, protein, fat, carbohydrates, and sugars. There are many minerals also found in this wondrous syrup. It has calcium, iron, magnesium, phosphorus, sodium, potassium, and zinc, not to mention vitamins like thiamin, riboflavin, niacin, and B6. Yep, all of that's found in maple syrup.

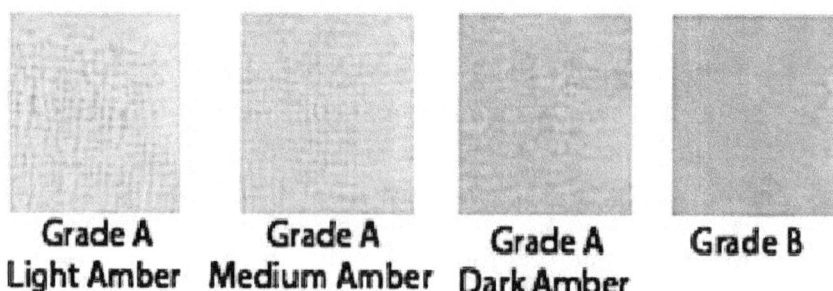

| Grade A Light Amber | Grade A Medium Amber | Grade A Dark Amber | Grade B |

The light amber color is made early in the season and has a very delightful flavor. Medium and dark amber have more of a rich flavor, and both are the most popular grades. The Grade B is a little harder to find and is gaining popularity in recent years. It has a very rich maple flavor and is widely used in baking and cooking, but is wonderful on pancakes. Make sure when you're buying your syrup that it is 100% maple with no blends with anything, including table syrups. This will ensure that you get the full health benefits of the maple syrup.

Organic Coconut Palm Sugar (also known as coco sugar, coco saps sugar, and coconut palm sugar)

We know that coconut has some very powerful properties. Is the sugar from coconut just as healthy? It has a very deep caramel flavor with a hint of butterscotch taste. It's made by making a cut to the coconut tree flower, the sap is collected in a large container and placed over a low heat for drying the juice. The juice is then condensed into brown sugar, which is one of the lowest glycemic index sweeteners that's available on the market. It rates in as 35-45 on the glycemic index, while other forms of sugar rate in at 65-100. Organic coconut palm sugar is very nutritious and makes the perfect choice for health conscious people without skimping on the taste.

Coconut sugar has been used in Southeast Asia for hundreds of years as part of their culinary heritage and is used in herbal medicine. If you're like many people concerned about your weight, this is a perfect match. With its ability to regulate the glucose and lipid levels in the body, this is a good food for people who suffer from both type 1-2 diabetes.

Have you ever had a sugar rush then the crash afterwards? This slow-energy-release sugar will keep you going all day long without having the jitters and the ups and downs of many other sugars. It is made up of 70-79% glucose and fructose 3-9%. Coconut sugar is high in potassium, magnesium, zinc, and iron, and is a natural source of the vitamins B1, B2, B3, B6, and C.

If you're going to use coconut as a sweetener, keep in mind that if it's not 100% pure, it might be mixed with cane sugar or malt-based ingredients. This may have an effect on the glycemic index.

Agave Syrup

The popularity of this syrup has gained ground since the early 2000s, and it can be found in many "health" food stores. This is primarily made of fructose **(agave syrup can contain in excess of 90% fructose)**, so it is similar to high fructose corn syrup. It has been endorsed by many celebrities and fitness personalities, who state this is a very healthy alternative to sugar. This sweet nectar was once used by the Mayans.

The high amounts of fructose can prevent the production of leptin. This hormone is responsible for making you feel full. Because you don't produce that hormone at adequate levels, you will eat more, causing weight gain. Just like high fructose corn syrup, the agave nectar is linked to insulin resistance and type 2 diabetes. It blocks the receptors for insulin, and the result is that your body makes more insulin to handle the glucose and your blood sugar level is harder to maintain.

Fructose stores in the body as visceral fat, or fat in the organs, manly the liver. This can lead to cirrhosis. Research has also found that agave has chemicals called anordin and dinordin. These chemicals are a kind of steroid that have a mild contraceptive effect that has been linked to miscarriages, even though miscarriages rare and you would have to consume a lot of agave. In this case, organic agave is not better for you. The agave plant naturally contains these harmful chemicals, and the organic version is no different. There is no nutritional benefit to agave nectar, and in fact, it reduces the important minerals in the body, including iron, magnesium, calcium, and zinc.

Organic Turbinado Sugar

Turbinado sugar is the more natural state of cane sugar and maintains much of its flavor. It has more moisture that of cane sugar because it's less processed, and maintains many of its nutrients. Turbinado sugar has 11 calories per teaspoon, while conventional white table sugar has 16 calories per teaspoon. Turbinado sugar keeps much of its molasses, and thus the vitamins found in sugar cane juice remains. It's estimated that 100 g of turbinado has about 85 mg of calcium, 100 mg of potassium, and 23 mg of magnesium, there are also small amounts of iron and phosphorus.

Sorghum Syrup (also known as milo or kafir)

Sorghum was first grown in Africa and was then brought to Asia and Europe before it was used in North America. There are four varieties of sorghum plant. These include grain sorghum, sweet sorghum, grass sorghum, and broom corn sorghum. If you're on a gluten-free diet or have celiac disease, sorghum syrup may be a good choice for you, as it's naturally gluten free. The nutritional value of sorghum varies based on the soil where it's grown, but here is the breakdown—about 70% starch, protein, amino acids, and vitamins like niacin, riboflavin, and thiamin. Sorghum also has very high levels of magnesium, iron, copper, calcium, phosphorus, and potassium, Sorghum has half the daily value of protein and fiber. In the last few years, sorghum has been grown from GMO seed. If you're going to be using sorghum in your diet, buy only 100% organic.

Natural Zero Calorie Sweeteners

These sweeteners have few to no calories. They were not as popular until recent years, even though they have no effect on your teeth and have little to no glycemic index. How safe are they, and what are they really doing to your health?

Luo Han Guo

Originally from the foggy mountains of southern China, this has been used for hundreds of years in China and Southeast Asia. Not only is it sweet, it even has medicinal properties. Luo Han Guo is used in soup and teas to treat symptoms of colds, coughs, sore throats, gastrointestinal disorders, and as a blood purifier. Not much research has been conducted, but at Nihon University in Japan, their research found nearly twenty compounds from this fruit with the potential to slow tumor growth.

It's also a great source for vitamin C, protein with 18 kinds of amino acids, and it's loaded with antioxidants. Luo Han Guo fruit is even used as an anti-inflammatory, tt can be found in Asian stores, but make sure that it's 100% the real thing. If you can find the whole fruit, that's a big plus, it's normally dried.

Stevia Bertoni (Stevia, Stevioside, Sweet Vibes, SweetLeaf, Honey Leaf)

This sweetener is derivative of a shrub that's native to Paraguay. It was used for hundreds of years by the Guarani Indians. They used it to add sweetness to their herbal tea, and it was harvested for the first time as a crop in 1908. Did you know that the FDA has deemed stevia as an unsafe food additive for over twenty years? They even at one point went as far as trying to stop the import of stevia in the United States. The import was accomplished under legislation passed in 1994. After a while, the makers of stevia were allowed to market stevia as a dietary supplement.

Research has suggested that the use of stevia puts you at risk of weight gain. Other risks may affect the male sex organs— this was the result of testing by a group of European scientists in early 2006. Animal studies done on lab rats that were fed high doses of stevia for twenty-two months.

The sperm production of the rats was severely reduced, and an increase of testicular cell proliferation. This could lead to infertility and reproductive problems, another study was done on female rats showing that when fed stevia, had smaller babies, and got pregnant less often. If you're planning on having children or are currently pregnant, don't eat stevia.

191

Stevia has also been linked to cancer in preliminary studies, as it can convert into mutagenic compound. This then can turn into cancer by the mutation of the cells. According to the Center for Science in the Public Interest, large amounts of this sweetener can inhibit the absorption of carbohydrates, as observed in animal studies. This can lead to lower energy levels and is of particular concern when it comes to children.

Stevia has side effects, and here are the more common ones:

- Dizziness

- Muscle pain

- Numbness

- Nausea

- Gas

- Bloating

Even though they are not that serious and only last for a few short hours, if you're having any of these side effects, it would be wise to contact your physician.

Thaumatin (also known as talin)

Thaumatin is made from the katemfer fruit. It's a naturally sweet protein, not a carbohydrate. This plant wasn't used in the western world until W.F. Daniell served in the British Army as a surgeon. He was stationed in West Africa in the early 1800s, Daniell notated it in his pharmaceutical journal. Thaumatin is about 2,000 to 3,000 times sweeter than normal table sugar—that's the most sweetness of any natural sweetener. The studies conducted did not find any adverse effects of thaumatin, even though the FDA has not approved as a food additive or sweetener.

Modified Sugars

These sweeteners are the product of man modifying the starch by using enzymes. One of the most well-known modified sugars is high fructose corn syrup.

Refiner's Syrup (also called golden syrup)

Refiner's syrup is made at a sugar refinery and is the byproduct of making white table sugar. It's devoid of all nutrients and is linked to all of the health effects that granulated sugar is linked to.

Inverted Sugar (or invert sugar)

This is a combination of glucose and fructose, which is accomplished by splicing sucrose into these two components. This makes a sold as a viscous liquid that is referred to as trimoline or invert syrup. It's sweeter and retains moisture.

Inverted sugar is commonly made from white table sugar that is why it has all of the same health effect as table sugar i.e. tooth decay, diabetes, and obesity. Inverted sugar is very similar to High Fructose Corn Syrup, the difference is that invert sugar is made from white table sugar and contains less fructose compared to HFCS. (Inverted sugar 37.5% vs. HFCS 55%.) Fructose has been linked to being a carcinogen in animal studies. In my opinion, the significant increase in the total amount of corn sugar, corn syrup, invert sugar, and sucrose added to foods would constitute a dietary hazard.

Artificial Sweeteners

Many artificial sweeteners have hit the store shelves in the past few years, and many have been known to cause some severe harm. These kinds of sweeteners have been used in the USA for over 120 years. Even though they don't affect your teeth, some have been linked to more dangerous conditions.

Sucralose (also known as Splenda)

Sucralose is sold as the brand Splenda, which was approved by the FDA in 1998. The FDA claims they have conducted about 100 studies in a span of twenty years, but the new independent studies are raising alarms. Splenda has been linked to migraines, according to a study published in the August, 2006 issue of **''Headache: Journal of Head and Face Pain.''** There was a study conducted by Y.F. Sasaki that was published in 2002 *'Journal of Mutation Research.* This study showed that long-term use of Splenda damages the DNA in the gastrointestinal region of lab mice.

To date, there have been about six human trials on the effects of Splenda. The trial that lasted the longest was only for three months. How can you tell after only three months if a product that have adverse effects on lab animals is safe for human use?

Did you know that 3.3%-7.2% of sucralose stays in your body? There are even some studies that suggest that sucralose can cause liver damage. There is more research that needs to be conducted to clarify if that's the case, and truly see what the long-term effects are and if it is passed to your unborn child during pregnancy, because Splenda damaging the DNA of gastrointestinal organs.

Neotame

Neotame is the name for a new sweetener developed by NutraSweet (the original manufacturer of aspartame). Neotame is reported to be 13,000 times sweeter than table sugar, and about 30 times sweeter than aspartame.

Like aspartame, some concerns include gradual neurotoxic and immunotoxic damage from the compounding of the formaldehyde metabolite (which is toxic at extremely low doses) and the excitotoxic amino acid. If that's not bad enough, the FDA bent the rules on what is allowed in organic food. This was done to get neotame on the tables of the consumer, according to a recent article on Sott.net:

"Neotame was approved by the FDA for general use in July 2002… The FDA loosened all labeling requirements for neotame as part of a large-scale effort to make it a near-ubiquitous artificial sweetener, to be found on the tabletop, in all prepared foods, even in organics. It simply doesn't have to be included in the ingredient list."

Neotame has been made acceptable, and without being included on the list of ingredients, for:

1.USDA Certified organic food items.

2.Certified kosher products with the official letter k inside the circle on labels.

Products that already contain neotame:

- Kroger's fruit juice and certain powdered iced teas
- Detour energy bars (certain varieties)
- Roman Meal bread line
- Herr's pretzels

Acesulfame K

Acesulfame potassium is also known as Acesulfame K, Ace-K. This has 200 times the sweetness of normal table sugar while being very low in calories. The German chemist Karl Clauss accidentally discovered this way back in 1967. Sounds like the perfect substitute for sugar, right? Well, not really, even though the FDA approved it as a sweetener in specific foods in 1988, and for general purpose in the year 2002. There is substantial evidence that sparks criticism, even from the Center for Science in the Public Interest (CSPI). They don't approve of the use of this chemical, and according to the research, Ace-K may possess carcinogenic properties. The animal test that was conducted showed that Acesulfame K is linked to certain kinds of tumors. This includes lung tumors and breast tumors. It's even suggested that this chemical can cause various types of leukemia and chronic respiratory disease. The animal studies were performed, but no human testing has been conducted, to my knowledge, and the results were ignored by the FDA.

The side effects are:

Headaches Depression

Nausea Mental confusion

Liver damage Kidney damage

Acesulfame K is commonly found in baked foods, sugar-free beverages, chewing gum, gelatin, and sodas. This chemical is commonly called Acesulfame K, Acesulfame Potassium, Ace-K, or Sunett.

Cyclamate (also known as Sodium N-cyclohexylsulfamate)

Cyclamate is a synthetic food additive and a sugar substitute that was synthesized in 1937, but wasn't used by the masses until later. There are two varieties, sodium cyclamate and calcium cyclamate, unlike some other sweeteners, it doesn't have a metallic aftertaste. Cyclamate is 30 to 50 times sweeter than table sugar, making it the lesser sweet of the synthetic sweeteners. Studies have liked it to causing cancer, this sweetener is not recommended for women who are pregnant or lactating.

Studies conducted early on showed that cyclamate can change in the intestines, transforming into cyclohexamine, which is a known carcinogen. A study published in Toxicological Sciences suggested that cyclamate is linked to testicular atrophy and stopped the growth of sperm in lab monkeys fed cyclamate over a long period of time. Keep in mind that the monkeys were fed 20-100 time more than the daily intake necessary to cause a human carcinogen.

The Truth about Sugar

White sugar is the most common and widely used sweetener in America. It's the crystallized form of sucrose, it's made from sugar beets and high fructose corn syrup, heavily refined cane sugar comprising of 99.96% sucrose.

In 2008, Monsanto's Roundup Ready GMO sugar beets revolutionized the sugar industry and have completely taken over the market. It's estimated that 95% of the sugar beets grown in the United States are genetically modified. Approximately 50% of white sugar is made from GMO sugar beets and not sugar cane at all. At this moment, there are no regulations to establish whether "sugar" or "sucrose" is derived from beet sugar, cane sugar, or a mix of both.

While Monsanto clearly plans to make its GM sugar beets farmers only option, experts said that the seed companies would have enough non-GMO seed if the court decided to follow up its ruling that GM sugar beets were approved illegally a moratorium. White table sugar as being linked to cause type 2 diabetes, obesity, and cardiovascular disease risks.

Studies have shown that long-term intake of refined white sugar can lead to the following conditions:

Tooth decay Osteoporosis

Gout Hyperactivity

Immunosuppression Hypertension

Inflammation Vision problems

Pancreatic damage Hypercholesterolemia

Multiple sclerosis Nutritional deficiencies

Cancer Kidney problems

Depression Schizophrenia

White sugar is also called granulated sugar and comes in many forms, including nib sugar, table sugar, and caster sugar. It's used in making confectioner's sugar, commonly known as powdered sugar, and is mixed with cornstarch to prevent clumping. The media has scrutinized sugars, stating that they cause diabetes and other health problems. In the 1980, three people out of 100,000 had diabetes. According to the CDC, "In 2012, 29.1 million Americans, or 9.3% of the population, had diabetes." The estimated projections are going to rise over the years. Dr.Mercola recommends **"eating unprocessed, organic and locally grown foods"** to enhance health. Organic honey and organic cane sugar have nutrients that other sweeteners don't have. They are more easily metabolized and utilized than other forms of sweeteners. American Diabetes Services are now stating, **"In general, you should try to avoid eating or drinking too many products with artificial sweeteners. Opt for those with natural sugar substitutes instead."**

Section 26
Oral Contraceptive and Hormone Therapies

Even though you wouldn't think to put this section in a shopping book, I feel that it's crucial because you're able to get hormone therapies from many pharmacies, and oral contraceptives. **"In the fifty years since its release, the birth control pill has become synonymous with women's liberation and has been thought of as some sort of miracle drug. But now it's making women sick,"** Ricki Lake said in a statement along with director, Abby Epstein **"Our goal with this film is to wake women up on the unexposed side effects of these powerful medications and the unforeseen consequences of repressing women's natural cycles."** -The film, called "Sweetening the Pill," is still in production so as of yet to announce a release date, the book is available now where books are sold.

Women who take the contraceptive pill are two times more likely to get cervical cancer and have up to 30% more of a chance to develop breast cancer. Risks last as long as ten years after the woman's last pill, it's estimated that 150 million women globally are using the pill. Each pill contains high levels of the hormone estrogen, estrogen tricks your body into thinking it's pregnant, forcing your pituitary gland into making hormone, suppressing your period. The hormone makes your uterine lining thicken, when you take the placebo pills, your estrogen level drops suddenly, and your body menstruates normally.

Many women have reported some minor changes:

- Larger breasts

- Weight gain or loss

- Reduced or increased acne

- Slight nausea

199

- Emotional sensitivity right before their period
- Mood swings throughout their cycle
- Irregular bleeding or spotting
- Breast tenderness
- Decreased libido

These are common side effects when on the pill, and yet few doctors discuss the dangers of taking this drugs for long periods of time.

Here are some of the more serious symptoms of birth control:

- Increased risk of cervical and breast cancers
- Increased risk of heart attack and stroke
- Migraines
- Higher Blood pressure
- Gallbladder disease
- Infertility
- Benign liver tumors
- Decreased bone density
- Yeast overgrowth and infection
- Increased risk of blood clotting

One of the ever-growing causes of death among women is heart disease, which is now ranked as the leading cause. According to the CDC, one in four women will get heart disease.

Additionally, birth control kills good bacteria. That can lead to yeast infections, lowering your overall immunity and increasing the rate of infections.

Women who are on the pill before they get pregnant have a 40% higher chance of developing gestational diabetes than a woman who never used this form of birth control.

"This study provides evidence that hormonal contraceptive methods may increase a woman's risk for GDM (gestational diabetes) in her following pregnancy, even after adjusting for maternal age, race, education and income level, marital status, Medicaid status at delivery, and type of prenatal care received," the researchers wrote in the study released on 19 July, 2014, by the Centers for Disease Control and Prevention. In 2005, the World Health Organization officially classified oral contraceptives as Group I carcinogens (Group I, is the most dangerous from Group).

"The pill is a powerful endocrine disruptor with a whole body impact. It is one of the only drugs given to healthy people to take over a long period of time," says Grigg-Spall. **"It suppresses the endocrine, metabolic, and immune systems in every woman. It causes vitamin deficiency. It suppresses ovulation, which research shows has benefits outside of allowing for pregnancy. Consistent ovulation promotes bone, heart, and breast health long term and protects from a number of diseases that kill women at a high rate."**

In 2006, a meta-analysis published in the Mayo Clinic Proceedings stated that 21 out of 23 studies discovered an increased risk in breast cancer in women who had taken the pill before their first child. On average, the women experienced a 44% increased risk in developing breast cancer before age fifty. Even though the birth control pill is supposed to decrease the risk of other forms of cancer such as ovarian and uterine cancer, it has been shown to increase breast, liver, and cervical cancer.

About 1.5 million women and girls use this pill for non contraceptive purposes. One study, "Beyond Birth Control: The Overlooked Benefits of Oral Contraceptive Pills," by Rachel K. Jones of the Guttmacher Institute, found that about 58% of people use hormone therapies for purposes other than pregnancy prevention. One of the common reasons women are put on the pill is to reduce menstrual cramps or pain—this makes up about 31%, 28% is for treatment of acne, and 14% is treatment for conditions like endometriosis and PCOS. There are estimated to be about one million women who never had sex but use the pill for noncontraceptive reasons.

Why are doctors giving this drug for so-called "menstrual-related disorders" that are very common in young girls, and why are their parents not informed of the long-term damage that it could cause?

Why Estrogen

I started to notice a pattern when I was doing research for this book, the one thing that recurred over and over was estrogen. Many of the chemicals that are in foods, cosmetics, and plastics are estrogen mimickers, it is also well known that many pesticides block testosterone, but why is that a big deal? Some might say, "Great. I'm a woman, so this would make me more of a woman." Or one common reaction I hear is, "So I'll have a higher sex drive?"

First of all, let's have a look to see what the role of estrogen is in the body, and what function it fulfills. Estrogen is produced by both men and women, but women produce higher amounts than men. The main function of estrogen in females is the development of breasts, endometrium, regulation of the menstrual cycle, etc. In males, estrogen regulates the production of sperm and regulating libido, the full spectrum of estrogen in males is not entirely clear.

It's important to fully understand the function of the estrogen hormone. It is a vital chemical in humans and animals. It's a chemical messenger carrying information from one place to another. It controls every aspect of our cell structure that regulates our growth, development, metabolism, tissue function, sexual function, and reproduction. It even regulates the way our bodies absorb the nutrients from the food we eat.

The use of the term "estrogen" refers to all of the similar hormones in the body. There are three kinds of naturally occurring estrogens: estrone, estradiol, and estriol. In women, estrogen is predominantly produced in the ovaries. Estrogen can also be produced by fat cells and the adrenal gland. The conditions that can cause estrogen levels to drop are hypogonadism (in men) and polycystic ovarian syndrome (in women). Estrogen levels are also known to drop after childbirth. Too much estrogen is not a good thing. It promotes tumor growth, and is linked to many forms of cancer.

Research into "male menopause" conducted by the *New England Journal of Medicine* shows that this is triggered when both testosterone and estrogen are declining in the male body. This study of 400 men ages 20 to 50 showed that estrogen is very important for managing the fat in men, and testosterone is important for maintaining muscle mass. Combined, they regulate the sex drive in both men and women, but must be balanced.

The Dangers of too much Estrogen!

Estrogen is mostly considered a female hormone, even though this kind of steroid hormone is produced by both men and women. Most of the estrogen in women is made in the ovaries, and there are small amounts made in the testes, adrenal, and pituitary glands of males. Too much estrogen in men is not good. If you look around, how many men do you see with small man boobs, or a stomach like a woman who's four months pregnant? Estrogen can have negative effects on both men and women if it's found in high amounts.

Here are common symptoms of high estrogen in men:

Gynecomastia (also known as man boobs)

If estrogen is in abundance in the body, the cells in the breast grow, giving men "man boobs." This change now happens in half of young males around puberty, but this threat becomes more present around the age of 25-30. Estrogen balances testosterone, high levels of estrogen in men can lead to erectile dysfunction and little-to-no desire for sex.

Infertility

Men with high estrogen levels have a greater chance of infertility because estrogen lowers sperm mobility.

Other Risks include:

Stroke—estrogen is known for causing blood clots, giving a greater chance of a stroke.

Heart attack—older men make less testosterone, lower testosterone levels and higher level of estrogen can lead to cardiac disease.

Prostate problems— studies show that high levels of estrogen play role in the cause of prostate cancer.

Weight gain—this is a very common problem, one that's not easily solved. As you gain weight, you make more estrogen, which makes you gain more weight. Do you see the pattern and why it's hard to break?

One of the leading symptoms of excess estrogen is called estrogen dominance,starting in early puberty and affects both men and women, most men are not even aware that they can have estrogen dominance and that is why I focused on men first. Estrogen dominance is growing, and for many, this health risk should be of concern.

Diet is a big part of managing estrogen. Your diet can, in fact, be full of estrogen-mimicking ingredients. Having fiber in your diet can help reduce the amounts of unwanted excess estrogen, which can lower the estrogen in your blood, we all live in a world with feminizing chemicals known as xenoestrogens, chemicals like chlorine, which are added to meat and dairy products. The abundance of estrogen could be one of the reasons for the decline of fertility and the need for IVF procedures. Try eating organic foods whenever possible. This will limit your exposure to harmful xenoestrogens. One medical practitioner who practices alternative medicine stated that **"the problems with**

estrogen dominance are many. Two of the main ones are [that] the sperm count is headed toward zero and [also] we have developed an obese population with all the complications of heart trouble, diabetes, etc."

Symptoms linked to estrogen dominance:

Weight gain (most commonly in the hips and thighs)	Sweet cravings
Reduced sex drive	Increased PMS symptoms
Water retention	Irregular menstrual cycles
Hot flashes	Migraines
Cervical cramping, this is more common for pregnant women.	Chronic fatigue
Allergies	Sinus infections
Irregular moods, mood swings, bouts of depression	Anxiety
Resistant weight loss	Cold hands and feet
Premature beginning of menstruation	Loss of hair
Insomnia	Polycystic Ovarian Syndrome
Infertility	Osteoporosis
Enlarged breasts in men	Broken capillaries, specifically in the stomach, inner arms, and breasts

Many symptoms women commonly experience from estrogen dominance is fibrocystic breast disease. This is one of the most common non cancerous breast conditions. It's estimated that more than 50% of all women will develop fibrocystic breasts at some point. Fibrocystic breast disease can happen to anyone at any age, but is more common in younger women of childbearing age. The lumps, known as "fibroids," are non-cancerous lumps, they are normally located on the upper breast or outer side of the breast close the armpit, they can become noticeable about a week or so before a woman's period is due to start. They normally go away soon after the period begins. Changes to look for include a feeling of denseness and bumpiness in the breasts.

Dr. Jonathan Wright noted that **"some women have cysts so painful, they can't lie on their stomachs or even be hugged without pain."** Women may also experience dull pain or an itching sensation in and around the nipples, or a sensation of fullness (which some women describe as "swollen").

Another symptom may be reduced thyroid function, also known as hypothyroidism. It's not uncommon to have an underactive thyroid if estrogen dominance is present, and the thyroid hormone is hindered with high estrogen levels affecting thyroid hormone production.

Symptoms of hypothyroidism include weight gain, fatigue, intolerance to cold, constipation, inability to concentrate, depression, short-term memory loss, and dry skin. Research shows that one major health risk of estrogen dominance is cancer! Cancer cells feed on estrogen, and too much can lead to cancer.

The National Cancer Institute, estrogen has a role to play in cancer cell development. On their website, they state, **"The most serious problem arises from the ability of estrogen to promote the proliferation of cells in the breast and uterus."**

One of the reasons may be that estrogen is used by the body in the creation of new cell growth in many parts of the body, such as in the breast. The National Cancer Institute states that **"Mutations also can occur spontaneously as a result of mistakes that are made when a cell duplicates its DNA molecules prior to cell division. Although estrogen doesn't appear to directly cause the DNA mutations that trigger the development of human cancer, estrogen does stimulate cell proliferation."**

We live in a world that literally has thousands of chemicals that mimic estrogen. These exist in the food, water, air, plastic, and dozens of other environmental elements, mainly from old chemicals such as DDT that changes into DDE, another estrogen mimicker. Other examples are pesticides, dioxin, and PCBs (polychlorinated biphenyls), plus the one that we talked about in this book and many more that we might not know about.

Over the years, there have been more and more hermaphroditic children born every year, while the numbers of GLB (Gay, Lesbian, Bisexual) are on the rise. This is not by choice. I have spoken to my gay friends to get an understanding of their situation. They tell me that they feel like they are in the wrong sex's body. After researching this book, I have formed the opinion that estrogen and other genetic-modifying chemicals may be causing the hormonal alterations in humans like the pesticide Atrazine. **"Atrazine is the world's most widely used pesticides, wreaks havoc with the sex lives of adult male frogs, emasculating three-quarters of them and turning one in 10 into females, according to a new study by University of California, Berkeley, biologists."**

As the media has advertised unhealthy food as "healthy," it's no wonder why the body's hormone and immune system is haphazard. The new and ongoing cure is the "pill"—the pill for depression, a pill for being overweight, and a pill for almost anything. This is just wrong! Doctors are not treating the underlying problem—they're treating the symptom, and that pill then makes a new symptom to treat. This is an ongoing roller coaster, of illness and health, being medicated and then told you're still sick. There has never before been this much cancer and new forms of disease as there is now, almost every year, there are some new viruses or new medical conditions that requires medication to manage.

Section 27
How to Treat Estrogen Dominance Naturally

Disclaimer: I am no doctor, the doses are only a guide, you should consult your physician for the right prescribed amounts for you.

Progesterone is the natural rival to estrogen, and helps keep estrogen in check. Balancing the imbalance between estrogen and progesterone will treat many of these symptoms within a few months. Some women may find their estrogen dominance symptoms worsening when they start the progesterone treatment, such behavior is very common; It is because of the amount of stored estrogen in the body. To keep the symptoms from getting worse, it's suggested that women may want to start off with about 200 mg per day. If that doesn't help, don't stop. Push the dose up to about 400 mg per day and for even greater success, you shouldn't start off with less than 100 mg/ml of progesterone. Of course you can consult your doctor about recommended doses. One big mistake people often make is using too little. Progesterone levels drop over the course of about 13 hours, and applying progesterone, two times a day is required. After you start to notice a balance of your symptoms (normally over a few weeks or months), slowly reduce the dose of progesterone over several weeks to a level that fits your body and maintains the balance.

As estrogen is present in both men and women, so is progesterone, the treatment is the same. Males may experience some worsening of symptoms as well—just push through it. Males should start off with no lower of a dose than 6-12 mg of progesterone cream per day. If needed, you can increase to about 10-20mg per day. If you have very bad symptoms and the estrogen levels are very high, it might be necessary to use up to 100mg per day, of course please consult your doctor for their recommendations.

Eat organic meat, fish, and poultry. They are not full of hormones, antibiotics, or any other GMO contaminants. This is also true for organic vegetables and fruit—they are grown without pesticides, herbicides, or any other GM contaminants.

New York banned trans fats, and you should too. Trans fats increase the production of estrogen. BPA is an estrogen mimicker, so use glass, good quality stainless steel, or ceramics to store food and drink as much as possible. The use of birth control pills, condoms, and spermicides all have an effect on estrogen levels. Sometimes taking estrogen supplements can make you feel better at first, but this will only last for a short time. This is the reason why women are changing their Hormone replacement therapy (HRT) prescription often, as talked about in the previous section. The risk is not worth using the HRT or the contraceptive pill for any length of time.

Section 28
Behind the Mask

Some people think that hybridization, also called crossbreeding, is GMO, but there is a difference. Hybridization is taking two similar plants and pollinating them to create a new species, this process can take 6-10 generations. On the other hand, genetic modification is the process in which a DNA is forced into a plant or animal species that it doesn't naturally belong to. A prime example of the GMO's impact on the environment is the killer bees, some Brazilian scientists in the 50's were trying to create a bee that would produce more honey. The scientists took the DNA from a European honey-bee and mixed it with that of an African bee. The bees escaped from the lab and wreaked havoc on North and South America and even parts of Canada. Killing hundreds of people and animals since they escaped from the lab, they are ever growing in numbers and territory. If a bee can cause this much damage and death how would hundreds of GMO animals and crops change the world?

Genetic engineering forces the gene from a species, such as a spider, for example, and puts it into a goat. As they are two very different species, you can see the problem. GMO foods, such as corn, are genetically modified with a soil bacteria called Bt. This makes the corn immune to Roundup, herbicide, and E. Cali.

Other forms of genetically modified fruits and veggies include strawberries and tomatoes. Both have been spliced with a fish gene that's supposed to protect them from freezing. Some salmon have been genetically engineered with a growth hormone that makes them grow larger and faster than normal salmon. Cows are genetically engineered with a hormone rBGH or rBST to increase milk production and growth. Rice has been spliced with human genes to produce human serum albumin, some of the uses for human serum albumin are drugs and vaccine production, some vaccines contain genetically engineered human protein.

GMO foods are known as transgenic organisms. This means the foreign gene have been deliberately inserted into its genome could even affect the animals and humans who consume them, passing the genes to their unborn child. They have found Bt bacteria in the blood of women who are pregnant and in the blood of the unborn fetuses. Genetic engineering can result in both known, unknown, and unintended consequences.

The manufacturers of GMO food would have you believe that GMO foods would feed the world, and that the growing population needs this so-called wonder of modern science. Again, this could not be any further from the truth.

A thirty-year side-by-side comparison of GMO crops vs. organic shows that:

- Organic yields match conventional yields

- Organic outperforms conventional in years of drought

- Organic farming systems build rather than deplete soil organic matter, making it a more sustainable system

- Organic farming uses 45% less energy and is more efficient

- Conventional systems produce 40% more greenhouse gases

- Organic farming systems are more profitable than conventional

The research above was conducted by Rodale Institute Farming Systems Trial (FST), and was the longest-running side-by-side comparison of organic and chemical agriculture in the United States. The big myth about GMO foods is that they don't need as many herbicides and pesticides, but many of the GMO crops have the herbicides and pesticides spliced in their genes.

The plants have been changed to perform the following:

• To produce their own internal pesticide to kill or deter insects

• To remain alive when repeatedly sprayed with weed killers that are manufactured by these same corporations, including glyphosate (aka Roundup), glufosinate, and 2,-D (one of the primary ingredients of Agent Orange).

One of the common arguments about GMO foods is that "humans have been genetically engineering species for hundreds of years." The result is farming plants, a large variety of dog breeds, and seedless fruits such as grapes and watermelon. What people need to understand is that's not achieved by altering the genetics, but by finding ways to enhance the genes from hybridization. Hybridization is not genetic engineering!

March Against Monsanto in San Diego, California, May 2013

Many states are pulling together to ban GMO. Maryland banned genetically engineered fish. North Dakota and Montana placed bans on genetically engineered wheat. Burlington, Vermont, and Boulder, Colorado banned genetically engineered food, and GM crops have been banned in the California counties of

Mendocino, Trinity, and Marin, the big island of Hawaii, and the San Juan county of Washington state.

"The hope of the industry is that over time, the market is so flooded [with GMOs] that there's nothing you can do about it. You just sort of surrender." Don Westfall, biotech industry consultant and vice-president of Promar International, quoted in, "Starlink fallout could cost billions, *Toronto Star,* 9 January 2001.

As of early 2014, the USDA released the reports for the percentage of GMO crops grown in the U.S.

- Soybeans 94%

- Corn 93%

- Cotton 96%

- Sugar Beets 90%

The following represent percentages from 2012, as no current data can be found on the USDA site:

- Canola - 88%

- Hawaiian papaya - more than 50%

- Zucchini and yellow Squash - so small that it is not listed

- Quest brand tobacco - 100%

- Alfalfa (this was approved recently by the FDA; used as animal feed)

- Kentucky bluegrass (this was approved recently by the FDA; used as animal feed)

- Farmed salmon (using a growth hormone)

- Because the corn has been genetically engineered to produce the insecticide within it, the Environmental Protection Agency now regulates corn as an insecticide.

The problem is not the total amount of GMO crops in the U.S.—it's about the byproducts of these crops. They are used in almost everything from infant formula to bread, tofu, and tomato sauce. Farmers also use GM in the feed for their livestock, and this in turn gets into milk, eggs, and meat. GM can be found in products like ice cream, mayonnaise, cheese, veggie burgers. and in non foods like cosmetics, soaps, detergents, shampoo, and bubble bath.

The industry claims they have "thoroughly evaluated" the GM foods and have found them to be safe. This is just not true. There haven't been any long-term human studies conducted on GM food in the U.S. for over two decades. No American scientist has ever investigated the toxic levels of residues in GMO foods.

The irony is that some of the most powerful people in the world—who, in fact, support Monsanto—don't eat GM food. They only eat organic. *NaturalNews* states, **"While the masses, those very people who are supposedly representing and protected by their governments, are poisoned by hidden genetically modified organisms, pesticides and dangerous contaminants. The presidential family demands organic food in their kitchen, yet behind closed doors, shake hands with the biotech industry."**

Monsanto's own employees demand non-GMO food in their cafeteria. If these foods are so good, why would the makers of GM food and the leaders of the U.S. government not eat them? The First Lady Michelle Obama has gone on to talk about organic even when her husband is making deals and promoting GM and Monsanto. "You know, in my household, over the last year we have just shifted to organic," she said in a *New Yorker* interview during Barack Obama's 2008 presidential campaign.

Even though Monsanto's makes the claims that genetically modified organisms have never been proven to be a harm to human health. That's also not true—the main aspect of GM marketing is confusion. Based on numerous peer review studies, GM foods can cause many harmful effects, including grotesque tumors, premature death, organ failure, gastric lesions, liver damage, kidney damage, severe allergic reactions, and a viral gene that disrupts human functions.

Organizations like the FDA, the EPA, and the USDA all wear a shiny halo. They receive their power and influence from the mere fact that the public believes that their number-one priority is the health and safety of the citizens they are supposed to be serving. All of the agencies vow that they are there to protect the public on their websites.

The FDA:

> **"FDA is responsible for protecting the public health by ensuring the safety, efficacy, and security of human and veterinary drugs, biological products, medical devices, our nation's food supply, cosmetics, and products that emit radiation.**
>
> **"FDA is also responsible for advancing the public health by helping to speed innovations that make medicines more effective, safer, and more affordable and by helping the public get the accurate, science-based information they need to use medicines and foods to maintain and improve their health."**

The Vision Statement of the USDA:

> **"To expand economic opportunity through innovation, helping rural America to thrive; to promote agricultural production, sustainability that better nourishes Americans while also helping feed others throughout the world; and to preserve and conserve our Nation's natural resources through restoring forests, improved watersheds, and healthy private working lands."**

The EPA:

> **"The mission of the EPA is to protect human health and the environment. The EPA's purpose is to ensure that all Americans are protected from significant risks to**

human health and the environment where they live, learn and work."

Like every other company, the USDA, EPA and FDA all have marketing teams. They are there to make you feel good. Monsanto has been successful in court cases because the company has ties to the U.S. government. Some of Monsanto's executives held positions in the judicial and policy-making departments in the Bush, Clinton, and Obama administrations.

- Michael Taylor: VP of Monsanto—deputy commissioner of the FDA

- Roger Beachy: Director of the Danforth Plant Science Center (paid off by Monsanto)—director of the USDA National Institute of Food and Agriculture

- Elena Kagan: Obama Solicitor General (she took organic farmers vs. Roundup Ready Alfalfa case on Monsanto's side)—US Supreme Court justice.

- Clarence Thomas: General Counsel for Monsanto—U.S. Supreme Court justice.

- Margaret Miller: Monsanto supervisor—deputy director of Human Food Safety

- Donald Rumsfield: Board of Directors for Monsanto's Searle Pharmaceuticals—U.S. Secretary of Defense

- Ann Veneman: Monsanto Board of Directors—U.S. Secretary of Agriculture

- Linda Fisher: Assistant Administrator of the EPA—VP of Monsanto—deputy Administrator of the EPA

- Dr. Michael A. Friedman: Deputy Commissioner of the FDA—Senior VP of Monsanto

The commissioners, directors, and secretaries of these organizations are put there for a reason. I believe the reason is money—they control the food and medicine in all of the U.S., and parts of the world. By means of deception and misinformation, their combined efforts confuse the public with false accusations of health benefits endorsed by celebrities. Take everything from the FDA and EPA with a grain of salt—they have made strange claims about how radiation and pesticides are safe and acceptable in your food, but raw milk isn't safe or healthy for consumption.

Corporate bureaucracy would also have you believe that there is no difference nutritionally between genetically modified vs. organic. They will talk about the calories, the fiber, and all of the micronutrients, claiming that they are identical in every way as they laugh at people who they claim to pay "double to triple" the price to avoid GMOs in their diets.

Even world-renowned TV personality Dr. Oz stated on his show, **"Providing consumers with safe products is our number one priority, and we understand that some consumers have questions about genetically modified food ingredients. Genetically modified ingredients are not only safe for people and our planet, but also have a number of important benefits.**

"Virtually every credible food safety organization and scientific study have found genetically modified food ingredients are safe and there are no negative health effects associated with their use. GMO crops use less water and fewer pesticides, reduce the price of crops by 15-30% and can help us feed a growing global population of seven billion people." -The Grocery Manufacturers Association

In truth, there is a big difference in the nutrition of GMO and organic foods. A report that was released on the blog "Moms Across America" revealed the true nutritional differences. The report only talked about corn, but you can see a clear difference. The results are from the organic corn producer in Canada, De Dell.

• GMO Corn has 14 ppm of Calcium and Organic corn has 6130 ppm. 437 times more.

• GMO corn has 2 ppm of Magnesium and Organic corn has 113ppm. 56 times more.

• GMO corn has 2 ppm of Manganese and Organic corn has 14ppm. 7 times more.

The levels of glyphosate in the corn are not only scary, but the level should be classified as insecticide. The EPA standard level of glyphosate in water in the USA is 7 ppm. Studies have shown that organ damage begins at 1 ppb (. 0001 ppm). The GM corn contains up to 13 ppm, which is 130,000 times higher than what is known to be toxic in the water supply.

Studies conducted by Dr. Huber shown that that .97 ppm of formaldehyde is toxic if ingested by both humans and animals. GM corn has 200x that level, and animals are known to die after eating the corn crops.

There is more evidence of higher nutrient levels in non-GMO, organic foods:

A study published in the *Journal of Agricultural and Food Chemistry* confirmed that tomatoes grown by organic methods contain more phenolic compounds. A German study published in the Journal of Agricultural and Food Chemistry found that organically grown apples had a 15% higher antioxidant capacity than their conventional counterparts.

A review by the AFSSA (France's version of the FDA) concluded that **"organic plant products contain more dry matter and minerals—such as iron and magnesium—and more antioxidant polyphenols like phenols and salicylic acid."**

Source:responsibletechnology.org

The mouse on the right side of the photo was born to a mother that was only fed GM soy, as the mice on the left was fed a non-GM diet. As you can see, there is a clear size difference, and the mouse on the right has a higher mortality rate. The documented side effects of GMOs include immune dysregulation such as asthma, allergy, and inflammation; accelerated aging; infertility; dysregulation of genes associated with cholesterol synthesis, insulin regulation, cell signaling, and protein formation; damage to the liver, kidney, pancreas, spleen, and gastrointestinal system, stillbirth, birth defects, and early death.

20 April, 2011, the FDA granted Monsanto the ability to conduct their own internal studies on the impact on the environment. The Farm Bill gives Monsanto immunity from the FDA or USDA, and the bill also gives Monsanto control over where they can plant GMO test crops.

Norman Braksick, President of Monsanto subsidiary Asgrow Seed, stated back in 1994: **"If you put a label on**

genetically engineered food, you might as well put a skull and crossbones on it." (Source: *Kansas City Star*, March 7, 1994.)

What is the current status of GMOs? GMOs are causing a multitude of health issues, cancer rates are ever growing as the GMOs are on the rise. The USA has the sixth-highest number of new diagnoses, with 318 new cancer cases per 100,000 people. Cross-pollination is killing organic crops by causing a form of autoimmune response. This is bad for the environment because the organic plants don't have the chance to reproduce. The animals are being born with side effects such as infertility, or dying prematurely, because of, direct exposure to GM crops or even by consuming there kill. As pesticides have become more and more common in farming, the evidence shows that it is killing small microorganisms in the soil, so crops are ever more reliant on fossil fuel fertilizers.

"Cross-pollination of the environment is an issue, and that has to be addressed. And for those countries that have very small landmass, there's no way they can segregate GM crops from conventional crops or from organic crops, and so the likelihood of cross pollination exists," said Prof Patrick Wall, former Chairman of the European Food Safety Authority (EFSA), in an interview: **"We cannot force-feed EU citizens with GM food."** (2 December 2008)

How do the Big-Bio corporations respond to these problems?

The Threat to World Food Security
October 1998
by Ricarda A. Steinbrecher and Pat Roy Mooney
The Ecologist

"Monsanto's latest flagship technology makes a nonsense of its claim that it seeks to feed the world's hungry. On the contrary, it threatens to undermine the very basis of traditional agriculture—that of saving seeds from year to year. What's more, this 'gene

cocktail' will increase the risk that new toxins and allergens will make their way into the food chain.

"It was only in 1908 that George Shull came up with what Major Hallett really wanted—a biological weapon to keep farmers from saving and developing their own seeds. Called 'hybridization,' a wonderfully euphemistic term that led farmers to think that crossing two distant plant relatives could create a 'hybrid vigour' that so improved yield as to make the resulting seed sterility—meaning it could not be replanted—financially worthwhile. Today, almost every ear of corn grown from California to Kazakhstan is a hybrid controlled by any one of a handful of very large seed companies.

"Exactly 90 years after Shull's revelation, one of the biggest and most powerful of those companies, Monsanto, is fighting for control of the most important seed monopoly technology since the hybrid. But unlike 1860, this piece of life control can be patented. On March 3rd, the US Department of Agriculture (USDA) and a little-known cottoneed enterprise called Delta and Pine Land Company, acquired US patent 5,723,765—or the Technology Protection System (TPS).

Within days, the rest of the world knew TPS as Terminator Technology. Its declared goal is to promulgate plants that will produce self-terminating offspring - suicide seeds. Terminator Technology epitomizes what the genetic engineering of crops is all about and gives an insight into the driving forces behind the corporate campaign to control and own life.

"The Terminator rides to the rescue of long-suffering multinationals who have been unable to hold farmers back from their 12,000-year tradition of saving and breeding seeds. Farmers buy the seed once do their own work thereafter. Patents and Pinkerton detectives have

221

been employed to stop farmers from doing so. The Terminator though provides a built-in biological 'patent' enforced by engineered genes. Small farming communities of the Third World especially, rely upon their own plant breeding since neither corporate nor public breeders show much interest or aptitude in breeding for their often difficult environments. Old-fashioned hybrids and the Terminator Technology with its terminated seeds force farmers back to the market every season. Terminator also scuttles community conservation of agricultural biodiversity. There's nothing to conserve. It is the 'neutron bomb' of agriculture.

"The centuries-old practice of farmer-saved seed is really a gross disadvantage to Third World farmers who inadvertently become locked into obsolete varieties because of their taking the 'easy road' and not planting newer, more productive varieties." - Dr. Harry B. Collins, Delta and Pine Land Co, Vice President for Technology Transfer (June 12, 1998)

Monsanto and the biotech companies are notorious for telling the public that they are improving crop yield and nutrition to "feed the world." As you can see, their main focus is not that. They make 90% herbicide and pesticide resistance, increasing the toxicity in the food and making the public sick. They are forcing organic farms to spend more money on a 100-foot buffer zone around their fields, increasing the cost of good organic foods. If somehow the GM is found in organic fields, they sue the organic farmers for patent infringement and non-payment of licensing, royalty, and seed fees.

Oh, but they don't stop at the farms, the biotech companies such as Monsanto's have sued any state that tries to oppose them and requesting that GMO's should be labeled for the safety of the public. In mid-2014, President Barack Obama signed H.R. 933, which contained the Monsanto Protection Act, into law, even though more than 250,000 Americans asked to veto the bill.

Section 29
The Truth About Autism

This topic might be very controversial for some, you might even wonder why it's included in a book about shopping. I've decided to add this section for one very good reason—many shopping centers around the United States have pharmacies that offer flu vaccines. In recent news, there has been a lot of talk about vaccines and the dangers they pose to the public.

I don't mean to offend anyone—that's not my intention. I'm merely trying to help piece together the information and hope you'll understand that I'm only doing this for everyone to see what's going on under our noses. This subject is also close to me, as it is for thousands of people around the world—my little brother was diagnosed with Asperger's, so I started to research Asperger's in hopes of helping him to have a full and happy life.

What is autism really?

From the early 20th century, autism has been referred to as a range of neuropsychological conditions. Well, what does that mean? Let's break it down. The word "neuro" means "nerve" or "nervous system," and the word "psychological" is the way in which our minds process information in our lives. So, autism really means the way a person perceives their surroundings, and the brain's interpretations of that information.

The word "autism" is from the Greek word "autos," which means "self." "Autism" was first used in 1911 by a psychiatrist named Eugen Bleuler. He used the word to describe schizophrenia, which is a group of symptoms that include distorted thoughts, hallucinations, and feelings of fright and paranoia. In the 1940s, researchers in the United States started to use this term to describe children's emotional and social behavior around the same time a scientist from Germany, Hans Asperger, started to notice similar behaviors and medical conditions which are now known as Asperger's syndrome. Up until the 1960s, scientist believed that autism and schizophrenia were in some way linked. Later they realized there is a profound difference, and started to understand that difference.

Many treatments of autism were mainly focused on LSD types of drugs, electric shock, and trying to change the behavior by means of unproven techniques. Many of the so-called treatments were about inflicting pain/punishment in hopes of changing behavior. Currently, the treatment for autism is behavioral therapy.

The first documented case of autism was in the United States. His name was Donald Gray Triplett, or just "Case 1... Donald T." He was diagnosed in late 1930s or early 1940s. His case was described in a 1943 medical article that announced the discovery of a condition unlike "anything reported so far." In the same medical article, Case 2 was mentioned. Back in the 1940s, there were only three documented cases of autism in the USA.

From the 1940s, the numbers of autistic children have skyrocketed in the U.S. Here is a chart that shows the number of children from the 1940s to the early 2000s.

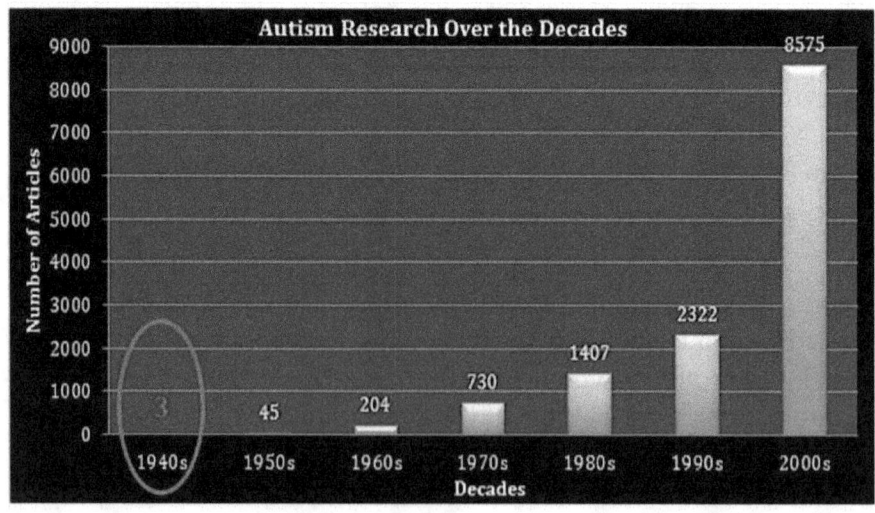

Source: readingroom.mindspec.org

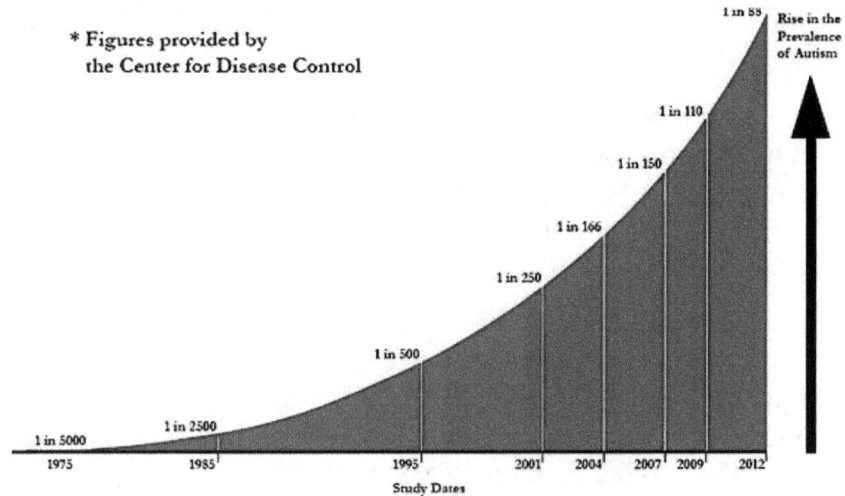

1 in 55 · Rise in the Prevalence of Autism

1 in 110

1 in 150

1 in 166

1 in 250

1 in 500

1 in 2500

1 in 5000

1975   1985   1995   2001   2004   2007 2009   2012

Study Dates

Source: carinoga.com

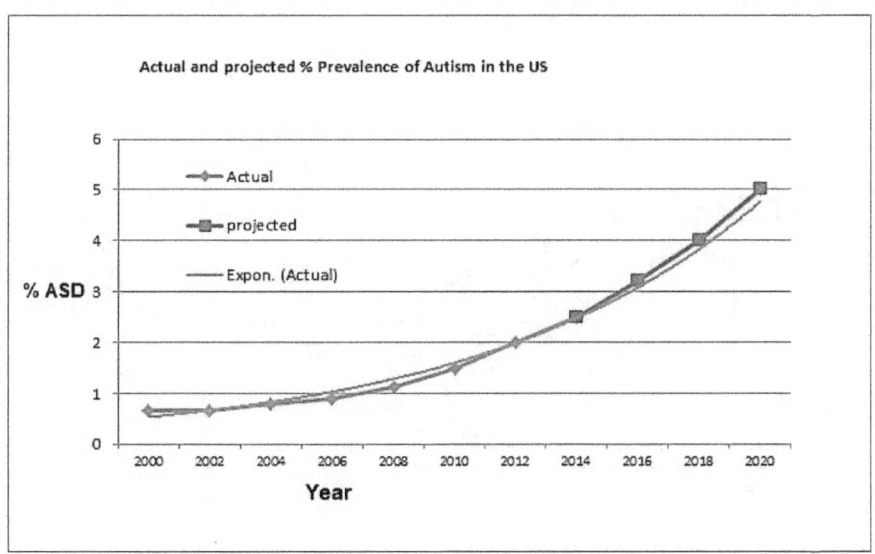

Actual and projected % Prevalence of Autism in the US

Year

Today, the rates of autism are now about 1 in 60. Why did they go from three cases to one in 60 in only 74 years? The U.S. government estimate that the rates of autism will increase about 3% every year. Let's have a look at what autism is doing to the body. Autism can start at any age, on average it starts in early childhood and can affect the central nervous system, the digestive system, and the immune system. Autism can manifest as early as infancy.

An infant with autism may be unresponsive to people or focus intently on one item to the exclusion of others for long periods of time. It's not uncommon for a child who manifests with autism later in life appear to develop normally and then withdraw and become uninterested in social engagement.

Many children who have autism appear to lack empathy. They might perform, repetitive movements such as rocking or twirling, or even banging their heads. They also tend to start speaking later than other children; they seem to have the inability to interact with other children in a normal manor.

Parts of the Brain Affected by Autism

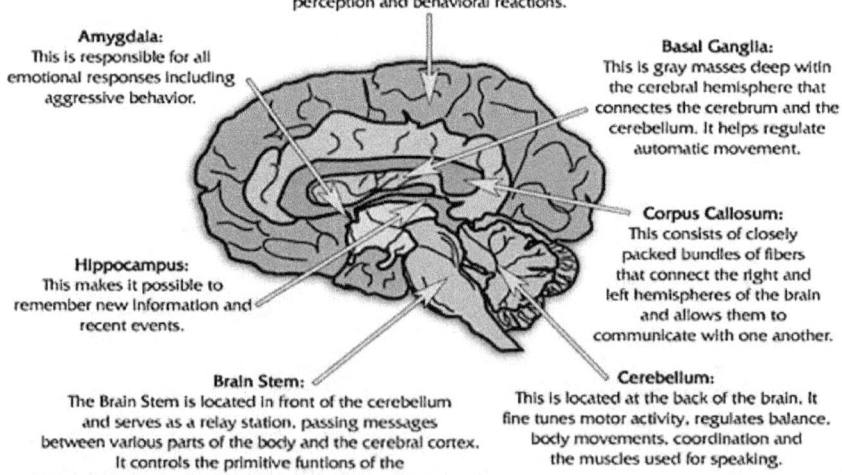

Cerebral Cortex:
A thin layer of gray matter on the surface of the cerebral hemispheres. Two thirds of this area is deep in the tissues and folds. This area of the brain is responsible for higher mental functions, general movement, perception and behavioral reactions.

Amygdala:
This is responsible for all emotional responses including aggressive behavior.

Basal Ganglia:
This is gray masses deep witin the cerebral hemisphere that connectes the cerebrum and the cerebellum. It helps regulate automatic movement.

Hippocampus:
This makes it possible to remember new information and recent events.

Corpus Callosum:
This consists of closely packed bundles of fibers that connect the right and left hemispheres of the brain and allows them to communicate with one another.

Brain Stem:
The Brain Stem is located in front of the cerebellum and serves as a relay station, passing messages between various parts of the body and the cerebral cortex. It controls the primitive funtions of the body essential to survival including breathing and heartt rate.

Cerebellum:
This is located at the back of the brain. It fine tunes motor activity, regulates balance, body movements, coordination and the muscles used for speaking.

Source: childrenshospital.org

Because autism doesn't manifest in every person about the same age, this leads me to believe that it is solely an environmental effect and not genetic. Not only does it affect children at different ages, but parents often describe it as an all-of-a-sudden change in their child that seems to have happened overnight. Some parents claim that they have "lost their child," and "they were not like that just a little while ago." From all the research I have done, autism affects the way the brain and body interpret information, or in other words, the way the neurons transmit messages with the help of electrochemical processes.

The United States has the highest rates of autism in the world. Why is that? What changed back in the 1940s? There are numerous speculations about what might be the cause of autism. In the past few sections, we had a look at some foods that scientists suggest could be linked to autism. But food alone wouldn't cause this epidemic. I believe that food, vaccines and other environmental factors are attacking us on a mass scale.

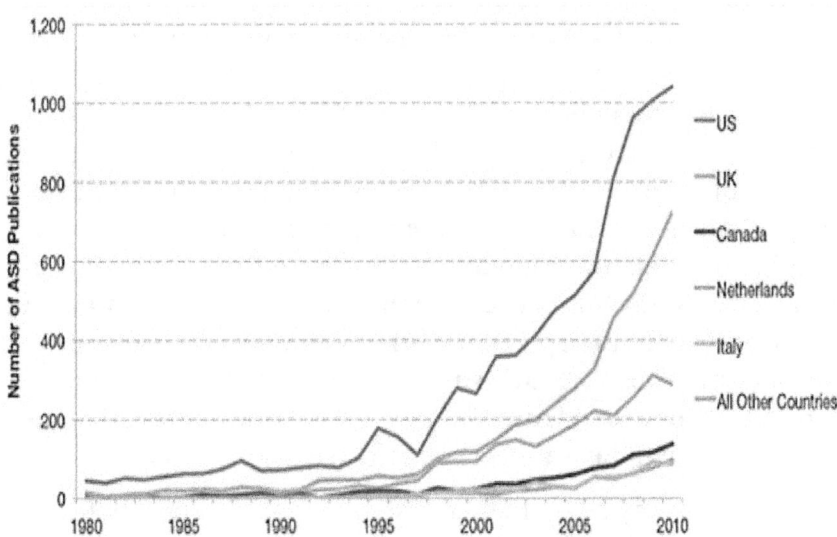

Source: iacc.hhs.gov

According to a study published on April 16th, 2012, the explosion of autism among children in the U.S. could very well be associated with the American diet. The CDC also stated that the rates are five times higher in males than in females. This study, according to Renee Dufault and his team, explores how the diet is affected by factors such as high fructose corn syrup (HFCS). It was released after the Centers for Disease Control and Prevention (CDC) estimated that an increase of 78% from 2002 through 2008.

Dr. David Wallinga, a physician at the Institute for Agriculture and Trade Policy (IATP) and co-author of the study, said: **"To better address the explosion of autism, it's critical we consider how unhealthy diets interfere with the body's ability to eliminate toxic chemicals, and ultimately our risk for developing long-term health problems like autism."**

The approach that Renee Dufault, a former FDA toxicologist, called "macro epigenetics." This describes the subtle side effects of HFCS consumption, including other dietary factors and their effect on the human body. The research could link the diet with chronic disorders.

The model they have developed factors in the nutritional and environmental factors as well as genetic makeup so they could see how they all might contribute to potential developments of certain health disorders. Dr. Richard Deth, a professor of Pharmacology at Northeastern University and a co-author of the study, explained: **"Rather than being independent sources of risk, factors like nutrition and exposure to toxic chemicals are cumulative and synergistic in their potential to disrupt normal development. These epigenetic effects can also be transmitted across generations. As autism rates continue to climb, it is imperative to incorporate this new epigenetic perspective into prevention, diagnosis and treatment strategies."**

There may be more than just food that can contribute to the development of the numerous neurodevelopmental disorders. More research must continue to analyze these effects from modern food and the environment and their relationship with autism spectrum disorder.

Even the substantial evidence from multiple scientific bodies states that GMO food are causing autism. Don Huber, PhD, a professor at Purdue University, is one of the leading men on the front lines when it comes to GMO crops and autism correlation. Huber spent fifty years in the field of botany (study of plants). Huber researched plant-borne pathogens that may be contained in the soil or the plant itself. In October of 2011, he discussed the physiological, neurological, and behavioral symptoms of pigs, cows, and rats been fed GM feed. After Huber's talk on GM food and livestock, other scientists approached him and stated, **"The symptoms you have described are exactly what we are finding in autistic children."** Huber started thinking, what if GM crops played some role in autism in children? There are some scientists who claim they see similar behavior in animals that are fed GM food. The animal's' offspring show an increase of the behavior as the GM foods are eaten over generations.

This raises the concern of many experts believe that the digestive problems from eating GMO foods might have a link to autism. Even though many people think GMO foods were not used until 1993-94, in fact, the genetic engineering of food has been around longer. Early scientists started using Gregor Mendel's genetic theory to manipulate and improve plant species in the 1800s. Mendel called this form of food modification "classic selection." This takes one kind of plant and mixes in other plants that are closely related to produce desired characteristics. Mendel is famous for his pea plant experiments, which were conducted between 1856 and 1863.

From these experiments, we discovered many of the rules of heredity, now known as the laws of Mendelian Inheritance. In 1953, when James Watson and Francis Crick published their findings of the helix structure of DNA, this discovery paved the way for modern scientists the ability to splice the genes from one organism into another. In 1973, Herbert Boyer and Stanley Cohen combined their research to create the first successful recombinant DNA organism.

GMO foods have been produced for a very long time. gout it have posable for some of the experimental plant pollen got out? The research shows that 70% of children who have been diagnosed with autism have digestive disorders. These symptoms include inflammation, intestinal permeability, and imbalance in the bacteria within the digestive tract. Animals fed GM show the same symptoms, Dr. Huber stated: **"When you look at the intestine on those pigs fed the GMO feed, the lining is deteriorated, and the critical microbial balance is drastically changed."**

There is mounting evidence that GMO food might have a part to play in the neurological disturbances that we are seeing in livestock and young children. According to research conducted by the U.S. Department of Veterans Affairs, **"Disruption of gastrointestinal flora by the use of antimicrobial agents or in relation to an immune defect, or poorly developed flora in young infants, may lead to proliferation of certain pathogenic microorganisms in one or more regions of the gastrointestinal tract. Production of toxins, particularly neurotoxins or toxic metabolic products, by such microorganisms may mediate neurological disruptions. It has been noted that certain neurological diseases have accompanying gastrointestinal manifestations, particularly constipation and diarrhea. This suggests the possibility that an intestinal microorganism may be the cause of both aspects of the disease."**

GMO' are not the only environmental toxin that can cause autism, vaccines are very controversial subject when it comes to the discussion. In August of 2014, Dr. William Thompson, senior scientist at the CDC, worked with other CDC scientists and top officials to hide data that established that black males are 340% more likely to be given an autism diagnosis when injected with MMR before the age of three. Dr. Thompson is a Ph.D., who worked for the CDC for over ten years. He is the author of a multitude of studies on the relationships between MMR vaccines and autism—now he is stating that the data was manipulated and hidden. NaturalNews received a letter from William Thompson to former CDC head Dr. Julie Gerberding, dated February 2nd, 2004.

This letter's date is of some importance because a critical Institute of Medicine (IoM) meeting regarding the safety of vaccines was to take place just one week later on February 9th.

In the letter, Dr. William Thompson stated: **"Presenting the summary of our results from the Metropolitan Atlanta Autism Case-Control Study, I will have to present several problematic results relating to statistical associations between the receipt of MMR vaccine and autism."**

(See the full letter on the next page.)

February 2nd, 2004

Dear Dr. Gerberding,

We've not met yet to discuss these matters, but I'm sure you're aware of the Institute of Medicine Meeting regarding vaccines and autism that will take place on February 9th. I will be presenting the summary of our results from the Metropolitan Atlanta Autism Case-Control Study and I will have to present several problematic results relating to statistical associations between the receipt of MMR vaccine and autism.

It's my understanding that you are aware of several news articles published over the past two weeks suggesting that Representative David Weldon is still waiting for a response from you regarding two letters he sent you regarding issues surrounding the integrity of your scientists in the National Immunization Program. I've repeatedly asked individuals in the NIP Office of the Directors Office why you haven't responded directly to the issues raised in those letters and I'm very disappointed with the answers I've received to date. In addition, I've repeatedly told individuals in the NIP OD over the last several years that they're doing a very poor job representing immunization safety issues and that we're losing the public relations war.

On Friday afternoon, January 30th, I presented the draft slides for IOM presentation to Dr. Steve Cochi and Dr. Melinda Wharton. The first thing I stated to both of them was my sincere concern regarding presenting this work to the Institute of Medicine if you had not replied to Representative Weldon's letters. I have attached the draft slides for your review. I have been told that you have suggested that the science speak for itself. In general I agree with that statement, but as you know, the science also needs advocates who can get the real scientific message out to the public.

In contrast to NIP's failure to be proactive in addressing immunization safety issues, you have done an amazingly effective job addressing the press on a wide range of controversial public health issues including SARS, Monkey Pox and Influenza. The CDC needs your leadership here because I may very well be presenting data before a hostile crowd of parents with autistic children who have been told not to trust the CDC. I believe it is your responsibility and duty to respond in writing to Representative Weldon's letters before the Institute of Medicine meeting and make those letters public. Otherwise, you give the appearance of agreeing with what he has been suggested in those correspondences and you're putting one of your own scientists in harms way. This is not the time for our leadership to act politically. It is a time for our leadership to stand by their scientists and do the right thing. Please assist me in this matter and respond to Representative Weldon's concerns in writing prior to my presentation on February 9th.

Sincerely,

William. W. Thompson, PhD
Epidemiologist
Immunization Safety Branch

233

This letter proves that even back in 2004, the CDC knew that vaccines were one of the causes of autism. One thing that needs to be discussed is infant immunization. **"Just before President Bush signed the homeland security bill into law, an unknown member of Congress inserted a provision into the legislation that blocks lawsuits against the maker of a controversial [mercury-based] vaccine preservative called 'thimerosal,' used in vaccines that are given to children."**— CBS, 12/12/02

The United States has the highest rates of child immunizations in the world. A study conducted in Germany was published on September of 2011 dealing with 8,000 children from ranging from newborn to nineteen years of age. The vaccinated children showed 2-5 times more diseases and medical disorders than the unvaccinated children. This chart shows the data collected and compared to the national German KIGGS health study. Something to note many of the survey's respondents came from the U.S.

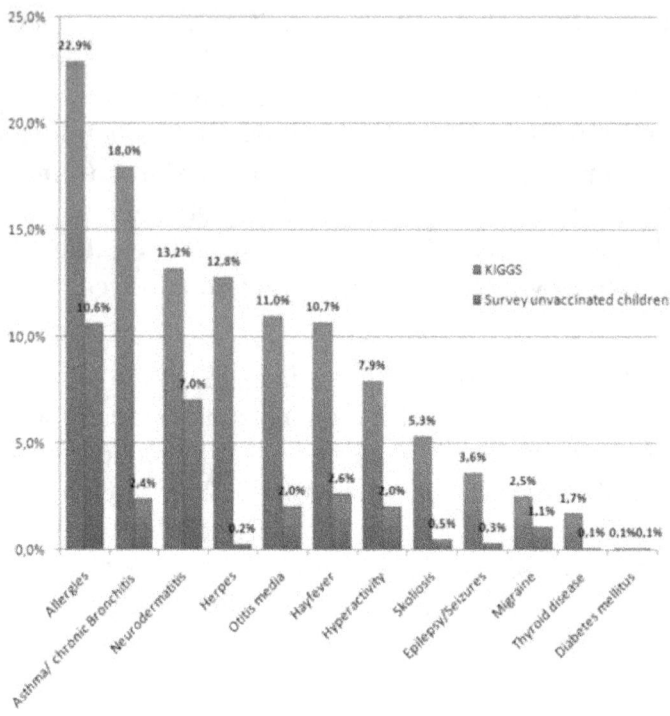

Source: vaccineinjury.info

The study is ongoing and will require more data, but even based off their current findings, the children who are immunized are more likely to become sicker in life, and vaccines are not helping at all. It was only a hundred years ago that children were given just one vaccine, and that was smallpox. About forty years ago, they got five vaccines: diphtheria, pertussis, tetanus, polio, and smallpox.

Back in the 1940's the child received 4 vaccines before age 2 Tetanus, Diphtheria, and Pertussis (DTP) and Smallpox and never more than one shot per doctor's visit. In 1980's a child would get 8 shots before age 2, and never more than 2 shots per doctor's visit. Now a child would receive 44 doses of 12-14 kinds of vaccines before the age of 2. the charts on the next page reflect 44 doses before age 2, some vaccines schedules show as much as 75 doses in total before the age of 6, this vaccination schedule for 2018.

Figure 1. Recommended Immunization Schedule for Children and Adolescents Aged 18 Years or Younger—United States, 2018.
(FOR THOSE WHO FALL BEHIND OR START LATE, SEE THE CATCH-UP SCHEDULE [FIGURE 2]).

These recommendations must be read with the footnotes that follow. For those who fall behind or start late, provide catch-up vaccination at the earliest opportunity as indicated by the green bars in Figure 1. To determine minimum intervals between doses, see the catch-up schedule (Figure 2). School entry and adolescent vaccine age groups are shaded in gray.

Vaccine	Birth	1 mo	2 mos	4 mos	6 mos	9 mos	12 mos	15 mos	18 mos	19-23 mos	2-3 yrs	4-6 yrs	7-10 yrs	11-12 yrs	13-15 yrs	16 yrs	17-18 yrs
Hepatitis B[1] (HepB)	1st dose	←─ 2nd dose ─→			←──────────── 3rd dose ────────────→												
Rotavirus[2] (RV) RV1 (2-dose series); RV5 (3-dose series)			1st dose	2nd dose	See footnote 2												
Diphtheria, tetanus, & acellular pertussis[3] (DTaP: <7 yrs)			1st dose	2nd dose	3rd dose		←──────────── 4th dose ────────────→					5th dose					
Haemophilus influenzae type b[4] (Hib)			1st dose	2nd dose	See footnote 4		←── 3rd or 4th dose, See footnote 4 ──→										
Pneumococcal conjugate[5] (PCV13)			1st dose	2nd dose	3rd dose		←──────────── 4th dose ────────────→										
Inactivated poliovirus[6] (IPV: <18 yrs)			1st dose	2nd dose	←──────────── 3rd dose ────────────→							4th dose					
Influenza[7] (IIV)						Annual vaccination (IIV) 1 or 2 doses						Annual vaccination (IIV) 1 dose only					
Measles, mumps, rubella[8] (MMR)					See footnote 8		←── 1st dose ──→					2nd dose					
Varicella[9] (VAR)							←── 1st dose ──→					2nd dose					
Hepatitis A[10] (HepA)							←── 2-dose series, See footnote 10 ──→										
Meningococcal[11,12] (MenACWY-D ≥9 mos; MenACWY-CRM ≥2 mos)						See footnote 11								1st dose		2nd dose	
Tetanus, diphtheria, & acellular pertussis[13] (Tdap: ≥7 yrs)														Tdap			
Human papillomavirus[14] (HPV)														See footnote 14			
Meningococcal B[11,12]															See footnote 12		
Pneumococcal polysaccharide[5] (PPSV23)															See footnote 5		

Range of recommended ages for all children

Range of recommended ages for catch-up immunization

Range of recommended ages for certain high-risk groups

Range of recommended ages that may receive vaccine, subject to individual clinical decision making

Range of recommended ages for non-high-risk groups

No recommendation

NOTE: The above recommendations must be read along with the footnotes of this schedule.

236

At the moment, there is no data that states that vaccines even help prevent any diseases, most vaccines are introduced at the last stage of the disease. I have never seen or heard of any research conducted on the multiple vaccines that are given in a short time span and what the long-term effect could be. Research is only available on single vaccines and only for the short term. Furthermore, there is no scientific evidence that the vaccinating of infants even has an effect.

As per senior doctors quoted by the *Times of India*, **"Children suffer from less than 2% of vaccine preventable illnesses, but 98% of the vaccines are targeted towards them."** There is one clear reason that the vaccines are being pushed onto the public, and that's money. Money from the vaccines themselves, and from the medical treatment from the lifelong effects from the vaccines.

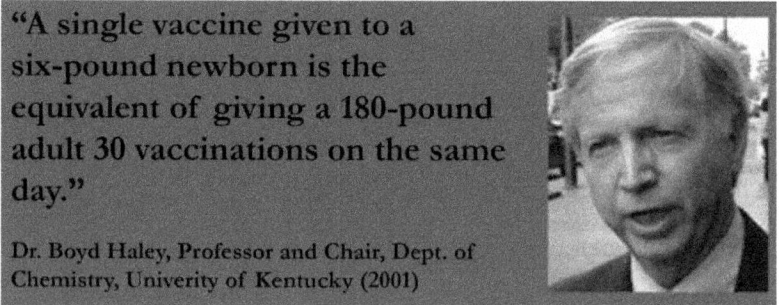

"A single vaccine given to a six-pound newborn is the equivalent of giving a 180-pound adult 30 vaccinations on the same day."

Dr. Boyd Haley, Professor and Chair, Dept. of Chemistry, Univerity of Kentucky (2001)

There's substantial evidence that points toward the fact that vaccines are not only harming children, but also are being forced on the general population in the United States. Still not convinced that vaccines are harmful? There are literally countless stories about infant deaths shortly after they have received their vaccines. Almost all of the vaccine-related deaths are not counted by the CDC as being from the vaccine—they are listed as SIDS (sudden infant death syndrome). Almost never does anyone ask about the infant's vaccination history.

According to the National Vaccine Information Center (NVIC): **"Hepatitis B vaccine-related adverse events reported to the federal Vaccine Adverse Events Reporting System (VAERS) [include] reports of headache, irritability, extreme fatigue, brain inflammation, convulsions, rheumatoid arthritis, optic neuritis, multiple sclerosis, lupus, Guillain-Barre Syndrome (GBS) and neuropathy. There have been more than 1,500 hepatitis B vaccine-related deaths reported, including deaths classified as sudden infant death syndrome (SIDS)."**

I strongly encourage anyone who has children or is thinking about having them in the future to do your research on vaccines and know what you're giving to your child. Are they really necessary? You know what's best for your child—not the doctor.

And remember, there is no law that states that you have to immunize your child, from the National Vaccine Information Center: **"Medical, religious, philosophical, conscientious or personal belief exemptions are worded differently in each state. To file and receive a vaccine exemption for your child to attend school, you must follow the regulations outlined in your state's vaccine law. In 2014, all 50 states allowed a medical vaccine exemption; 48 states allowed a religious vaccine exemption and 17 states allowed a philosophical, conscientious or personal belief exemption."**

Women who are pregnant shouldn't be getting any vaccines, even the flu vaccine that contains the deadly chemical thimerosal. I would like to make one thing very clear. I myself don't believe that GMO foods or even vaccines are the sole cause of increasing rates of autistic children that we are experiencing. I do, however, believe that all of them have a part to play in the rising rates of this pandemic, and until people force the makers of food and vaccines to change, we won't see a decline in the numbers of autistics children. All experts are in agreement that at this rate, we'll see an increase of about 3% per year.

Research conducted recently potentially links iodine deficiency and autism. The Centers for Disease Control and Prevention estimates that one out of every 60 children in the U.S. will display autistic symptoms. Scientists have proven the direct link between the underproduction of thyroxin in the thyroid and weakened neural connections in the brain. This comes from research conducted by James Adams at Arizona State University done by both urine and hair analyses on 51 autistic children, 29 mothers of autistic children, and a control group.

This study shown that iodine level in mothers might be the cause or intensifies the risk for autism. A study concluded that children with Autism spectrum disorder (ASD) had up to 45% lower levels of iodine than the control children. Based on an NHANES survey, the levels of iodine in U.S. residents are 50% lower than what they were between 1988-1994.

"It is increasingly apparent to us that autism is caused by environmental factors in most cases, not by genetics," said lead author Gustavo Román, M.D., a neurologist and neuro epidemiologist who directs the Nantz National Alzheimer Center. **"That gives me hope that prevention is possible."**

"Research has shown that mothers with low iodine levels early in their pregnancy are 4 times more likely to have a child with ASD," Dr. Gabriel Cousens says. **"Historically, as early as 1911, people normally took between 300,000-900,000 micrograms daily without incident. How is it that now only 1/5,000th of this dose is now considered safe? Even the Food and Nutritional Board at the Institute of Medicine has set the tolerable upper limit of 1,100 micrograms of iodine daily. Other researchers have used between 3,000 and 6,000 micrograms/day to prevent goiter. Iodine is found in every single one of our body's hundred trillion cells. Without adequate iodine levels, life is impossible. Iodine is the universal health nutrient and brings health on many levels."**

Dr. Gustavo Roman from the Houston Methodist Neurological Institute, in conjunction with researchers in the

Netherlands, directed studies of thousands of pregnant Dutch women. What they found was that the lack of sufficient iodine levels in mothers did in fact, affect fetal brain development. **"I think for the first time we have the possibility of finding an explanation of the problem, but most importantly, we have a way of preventing this from happening,"** says lead author Dr. Roman.

It is in my opinion that Autism spectrum disorder (ASD) is not the result of any one thing, nor is it the result of GMO or vaccines on their own. I strongly believe that autism is caused by the numerous harmful chemicals consumed by us for decades, dating back to the Industrial Revolution. Many of these chemicals are known to disrupt the iodine and estrogen levels in the body. At the same time, the use of GMO food and iodine deficiency are linked to gastrointestinal and neurological damage. It is also my belief that vaccines are not proven to prevent disease. I feel that vaccines add more chemicals, such as mercury, cause more brain and neurological damage. If you look at the information presented, you can see where autism is coming from. The research shows the same pattern over and over again. Get rid of GMO and the chemicals! Only use vaccines for the basics, take iodine, and autism rates will drop.

Section 30
Comparing the Cost

As I eat about 85-90 percent organic foods, the one comment that I get all the time is that organic food is so much more expensive than conventional food. The truth is, no, conventional foods are not cheaper. There are factors that make organic food cheaper, and in this section, we'll look at the cost for both conventional and organic food and compare them. Below is a generic shopping list. We are going to look at the prices from many organic shops and conventional stores.

Organic	Conventional	Difference
5lb Organic Russet Rio $1.12 per pound	Russet Rio $0.94 per pound	$0.18
Organic Roma Tomatoes $2.99 per pound	Roma Tomatoes $2.19 per pound	$0.81
Organic Cucumber $1.69 each	Cucumber $1.20 each	$0.49
Organic Lemons $1.18 per pound	Lemons per pound $2.19	$1.01
Organic Spring Mix Salad 16oz $5.60	Dole Spring Mix Salad, 10oz $4.68	$0.92 Organic spring mix was not sold as a 10oz pack at the time I looked it up.
Organic Red Onions $0.79 each	Red Onions $1.27 each	$0.48
Organic Fresh Parsley $1.49 each	Parsley $1.09	$0.40
Organic Avocados $1.29 each	Avocados each $1.25	$0.04
Organic Mushrooms Whole baby Bella pre-packed 8oz $3.89	Mushrooms, baby Bella pre-Packed 8oz $3.09	$0.80
Organic Bag Baby Carrots $3.29 2lb bag	Baby Carrots 2lb bag $4.38	$1.09
Organic Bell Peppers $1.07 per pound	Bell Peppers $3.00 per pound	$1.93
Organic Ear of Corn $1.00 each	Organic Ear of Corn $0.80 each	$0.20
Organic Garlic per pound $7.39	Garlic per pound $4.28	$3.11
Organic Granny Smith Apples per pound $2.01	Granny Smith Apples per pound $2.63	$0.62
Organic Bananas $0.45 each	Bananas $0.32 each	$0.13

Organic Green Seedless Grapes $2.99 per pound	Green Seedless Grapes $6.58 per pound	$3.59
Organic Whole Milk $6.59 per gallon	ole Milk $4.69 per gallon	$2.01
Organic Butter $7.69 16oz	!Butter $6.59 16oz	$1.01
Organic Sour Cream $3.09 16oz	Sour Cream $2.89	$0.20
Organic Monterey Jack Cheese $4.39 8oz	!Monterey Jack Cheese $4.39 8oz	Same
Organic Plain Yogurt $4.39 32oz	lain Yogurt $3.29 32oz	$1.10
Organic Beef Loin !Porterhouse Steak $12.50 per pound	!Beef Loin porterhouse Steak $13.19 per pound	$0.69
Organic Ground Beef $9.49 1.5lb	Ground Beef $8.83 1.5lb	$0.66
Organic Chicken Breast $8.79 1lb	Chicken Brest $9.89 1lb	$1.10
Organic Apple Gate Black !Ham Deli Meat $7.69 7oz	peli Meat Apple Gate Black Ham Deli Meat $7.69 7oz	Same
Qmw.!;:>t Organic Bead Grains and Seeds $5.49 24oz	QrQ.W.!;:>t 100% Whole eat $3.49 24oz	$2.00
Spectrum Organic Mayo $6.49 16oz	!Best Foods Mayo $5.29 16.5oz	$1.20
Organic Olive Oil S8.49 16.9oz	California Olive Oil $8.49 16.9oz	Same
!Kettle Organic Chips $3.19 5oz	Kettle $3.19 9oz	Same with a 4oz difference
Organic Ice Cream $5.99 1.5 Quarts	!Brevers Ice Cream $6.59 1.5 Quart	$0.60
Organic Pasta $2.55 1lb	>asta $1.99 1lb	$0.56
Organic Pasta Sauce $2.89 5oz	asta Sauce $2.79 24oz	$0.10
Organic Brown Rice $4.99 30oz	!Brown Rice $4.79 32oz	$0.20

Organic Tortillas $2.79 10oz	Tortillas $2.49 10oz	$0.30
Organic Can Beans $1.25 15oz	Beans Can $1.09 15oz	$0.16
Organic Can Soup $2.50 14.5oz	Can Soup $2.50 14.5oz	Same
Organic Raw Sugar $4.39 32oz	Raw Sugar $3.49 32oz	$1.00
Organic Flour $5.29 5lb	Flour $4.79 5lb	$0.50
Organic Juice $3.00 59fl	Juice $3.00 59fl	Same
Organic Coffee $6.49 10oz	Coffee $6.49 10oz	Same
Organic Ketchup $3.09 20oz	Ketchup $2.99 20oz	$0.10
Organic Eggs $4.39 12 count	Eggs $3.09 12 count	$1.30
Organic Cookies $4.59 9oz	Cookies $2.00 8oz	$2.59
Organic Valley Cream $4.99 1 liter	Creamer $4.39 32oz about 0.94fl	$0.60
Organic Baby Formula $32.99 25.75oz	Similac Infant Formula $32.99 16oz	Same price with a 9.75oz difference
Earth's Best Organic Baby food Jar $0.85-$1.00 4-6oz	Baby Food Jar $1.39 2.5oz	$0.39-$0.54 1.5oz difference
Organic Baby food Cereal $3.89 8oz	Baby food Cereal $3.09	$0.80
Total= $223.56	Total= $211.75	Total Difference= $11.81

Please note that I only "purchased" one of each item. All the prices are not sales price and without coupons. These prices were good as of October 27th, 2014, and were taken from Walmart, Vons, and Wholefoods. One thing that people don't understand is that organic foods give you more food for the price. And $11.81 is really not that much of a difference, even if you're on a budget.

The other factor that I don't see many people talking about is how often you eat after you've just eaten. Organic foods don't have MSG in them, and are not full of fillers. They will keep you satiated much longer, meaning that you will eat less, and less often.

People often don't realize that the food they eat are suppressing the full feeling that is telling you are full. This is one of the reasons why people are overweight, tt's estimated that 70% of Americans at some point in the year will end up purchasing organic food. It's my belief that it's becoming more common for people to look for healthy foods, and the truth about GM foods is getting out. I strongly believe that this trend is going to increase as the prices of organic foods drop.

The standard approved list of ingredients that are prohibited in organic products are:

- Antibiotics
- Artificial growth hormones
- High fructose corn syrup
- Artificial dyes (made from coal tar and petrochemicals)
- Artificial sweeteners derived from chemicals
- Synthetically created chemical pesticide and fertilizers
- Genetically engineered proteins and ingredients
- Sewage sludge
- Irradiation

The lack of these added ingredients are what differentiates organic foods from their conventional counterparts, according to the United States Department of Agriculture, at the current moment, there is no definitive measurement of insecticidal toxins within GM crops, or the levels of growth hormones in livestock. There are also no current measures for 80% of the antibiotics that are used for chicken, pork, and beef.

When I talk about the cost of organic foods, people always ask, "Why is organic more?" Taxpayer-funded subsidies are used for the GM farms and increase the organic farms' overall cost. This is to encourage people not to buy the so-called high-priced organic foods, even though they are cheaper long term. Because conventional foods don't have a defined measurement for the toxins contained within them, the long-term health effects of GM foods could in reality end up costing a fortune in doctor bills.

These long-term health costs are not spoken about much, and are a really big part of the overall cost of organic vs. GM foods. Conventional food allows cheap, synthetic, and very controversial ingredients that other countries have banned. Organic farmers and producers are then left to prove they are safe.

Would you really want to eat food with links to:

- Food allergy reactions
- Cancer and degenerative diseases
- Infertility
- Viral and bacterial illness
- Super Viruses
- Antibiotic threat via milk
- Antibiotic threat via plants
- Resurgence of infectious diseases
- Birth defects and shorter life spans
- Interior toxins
- Lowered nutrition

Many of the GM crops grow the pesticides within them, so while you're eating your butter-smothered corn on the cob, you're consuming 1,000 times the amount of BT toxin, compared to the organic crops. Toxins in pesticides are known to cause neurological problems, cancer, infertility, nausea, vomiting, diarrhea, allergies and asthma, wheezing, rashes and other skin problems, ADHD, birth defects, and a lot more. This section is to help people understand that you can still eat good healthy organic food even on a budget.

Almost all stores now use coupons, even the organic ones, check the websites of the organic stores in your area for coupons and special promotions. As social media grows, more companies are turning to this outlet and offer special coupons and deals.

Some of the best organic coupon sites ones are: mambosprouts.com, savingnaturally.com, organicdeals.com, organicfoodcoupons.com, healthsavers.com, organicdealsandsteals.com. Also have a look at Earth Fare coupons that can be found by going to their website earthfare.com/savings-coupons/couponbook.

Just like Walmart, some organic stores will price match.

Often, meat and dairy are the most expensive part of shopping. As these products contain pesticides, antibiotics, and growth hormone, they are the most important when shopping organic. Try not to oversee the value in purchasing these items organic. I often separate them in meal proportions at home or by asking the butcher for certain portions to cut on overconsumption.

When buying fruits and veggies, try not to purchase the pre-washed. They can cost twice as much, and may have been washed in fluoridated water. If you're like me and you buy sugar and flour in bulk, take a measuring cup so you can buy in cups and not pounds. This is great if you bake a lot. By doing this, you won't only know the weight, but also how many cups are there too.

Another great way to save money on organic foods is to shop online. Websites like greenpolkadotbox.com has good value for the membership. Amazon also sells organic and non-GMO, and has lots of good brands to choose from. VitaCost is a good, low-cost website with a variety of organic foods and vitamins. If you're going to venture out in the great world of organic online shopping, check out retailmenot.com, and the honey app for you web browser you can find the honey app by visiting their website joinhoney.com. Both have online promotional codes and discounts for many of the online organic stores. Organic brands like Trader Joe's, Earth Fare, 365, ShopRite, Wegman's, Kroger, Publix, and

Harris Teeter all have to follow the same strict rules when it comes to non-GMO and the USDA organic seal, so choose the brand that fits your budget.

The United States government continues to push the use of GMO crops and monoculture-based crops. However, the United Nations is once again (first time back in 2004) urging the United States to return to a more sustainable, organic system. This was the main point of a publication from the UN Commission on Trade and Development (UNCTAD) titled, "Trade and Environment Review 2013: Wake Up Before it's Too Late." This has contributions of more than sixty experts from around the world. The findings once again stated that organic and small-scale farming is the answer for "feeding the world," not GMOs and monocultures.

According to this report, a reduction in the use of fertilizer and many other changes are necessary. The report went on to state, claims that global security may in fact be at risk as food prices continue to rise, which I find very funny because Monsanto claims the GM crops are to lower the cost of farming, ultimately lowering the cost of food around the world. **"This implies a rapid and significant shift from conventional, monoculture-based and high-external-input-dependent industrial production toward mosaics of sustainable, regenerative production systems that also considerably improve the productivity of small-scale farmers,"** the report concludes. As a living human being, the freedom to choose is a birthright.

My Own Testimonial of Organic foods for Health

I have experienced a great deal of benefit from organic diet, I also take iodine and other natural cures for my overall health.

Here is my story.

I moved overseas for about seven years. The countries where I lived on average 10-15% GMO food, I moved back to the United States in early 2013. After about a week, I noticed that I was sick all the time, I had almost no energy, and a constant bloated stomach with foul-smelling flatulence and diarrhea all the time. About three weeks after I was back home in Utah, I even broke out in a rash on my face.

My girlfriend asked me to see a doctor, so I went to a dermatologist. They told me that I suffered a condition called face psoriasis, a chronic skin condition that is genetic, so he claimed. The physician prescribed me a medicated shampoo containing ketoconazole, when I researched to learn more about what was is in this shampoo, what I found was shocking. The oral prescription of Nizoral (also known as ketoconazole) has been related to serious liver damage and to effect adrenal gland production. Even though this damage is from the oral tablets, I still had to wonder how safe the shampoo was.

I started to reach natural treatments for psoriasis. The information I found stated that this so-called genetic condition is caused by GMO foods—that's why I never had it until I came back to the USA. I talked to my girlfriend, and we slowly made the switch to organic foods over about six weeks. Within the first week of the organic diet, I no longer had stomach cramps, bloating, and diarrhea. It took three weeks for the skin rash to completely clear up. I was 185 lbs when I started the diet, and I lost 20 lbs in eight months without exercising.

The moment when I noticed that it really was the GM food causing my sickness was when I went on a ten-day vacation only eating GM foods. It did not take long for the bloating, stomach cramps, and diarrhea to come back—about two nights. On the fifth day of the vacation, the rash was back. Only after eating an organic diet for about a week my rash was gone, and I had no more stomach problems.

To follow blindly without seeking knowledge is foolish, if you wake up one day finding yourself without health, prosperity, or the knowledge of how you got there, the only person to blame is yourself. Throughout the world, people live in fear and ignorance. They have given up their rights—for what? So-called security and safety.

In the end, they have not only given up their rights, but also the rights of their children and grandchildren. If you have the acceptance of ignorance, then expect to be taken advantage of by both country and corporations. In my belief, individual has right—to be fully informed, and the obligation of the federal government and corporations to be honest and restore the individual's constitutional right for the truth!

-E. A. Hargrave

If you enjoyed this book or found it useful I'd be very grateful if you'd post a short review. Your support really does make a difference and I read all the reviews personally, so I can get your feedback and make this book even better.

References

Adams, Mike. June 5th, 2013, NaturalNews.com

Adams, Mike. "New Berry-Based Natural Sweetener." *NaturalNews*. N.p. 2008. Web.

Ballerson, Terri. "The Dangers of Sunscreen." *Healthguidance*.org. Web.

Barret, Mike. "Fluoride Already Shown to Cause 10,000 Cancer Deaths." *Natural Society*. N.p. 2012. Web.

Barrett, Mike. "Toxic Metals Arsenic and Cadmium Found in Baby Food." *Natural Society*. N.p. 2011. Web.

Benson, Jonathan. "Spike in US Autism Rates Linked to High-Fructose Corn Syrup Consumption." *NaturalNews*. N.p. 2012. Web.

Benson, Jonathan. "Use These Five Natural Supplements to Detox Your Body of Toxic GMO Foods." *NaturalNews*. N.p. 2013. Web.

Boseley, Sarah. "Worldwide Cancer Cases Expected to Soar By 70% Over Next 20 Years." *The Guardian*. N.p. 2014. Web.

Bosque, Tomás. "The True Ingredients of Bottled Water | Ban the Bottle." *Banthebottle*.net. N.p. 2010. Web.

Bosworth, Adam. "Is Your Soda Killing You? Why Mayor Bloomberg Got It Right." *The Huffington Post*. N.p. 2013. Web.

Botelho, Greg . "Recall of Nearly 9 Million Pounds of Meat Not Fully Inspected - CNN.Com." *CNN*. N.p. 2014. Web.

Breyer, Melissa. "Risks of Sunscreen: New Report." *Care2*.com. N.p. 2010. Web.

Castillo, Michelle. "Global Cancer Rates Expected to Hit 22 Million New Cases Per Year By 2030: WHO." *CBSNews*.com. N.p. 2014. Web.

Chan, Amanda. "Scientists Create 'Bullet-Proof Skin.'" *The Huffington Post*. N.p. 2011. Web

Cirelli, Cheryl. "Dangers of Stevia." *LoveToKnow*. Web.

Colliver, Victoria. "BPA-Free Plastics May Be Less Safe Than Those with Chemical." *SFGate*. N.p. 2014. Web.

Dr. David. M.d. "truth In Skincare: Ingredient Watch: Sodium Hydroxymethylglycinate." *Truthinskincare*.com. N.p. 2008. Web.

Dailykos.com. "Study Finds Pigs Fed with GMO Grains to Have Health Problems." N.p. 2013. Web.

Dansinger, MD, Michael. "Insulin Resistance Syndrome (Metabolic Syndrome) Symptoms, Treatments." *WebMD*.com. N.p. 2014. Web.

Das, Asha. "5 Unhealthy Cooking Oils to Avoid." *BoldSky*.com. N.p. 2013. Web.

Dean, Tommy. "USDA Issues Recall of 9 Million Pounds of Meat." V*egNews*.com. N.p. 2014. Web.

Devon, L.J. "BPA Makes Breast Cancer Tumors Resistant to Chemotherapy." *NaturalNews*. N.p. 2014. Web.

Dherbs. "The Dangers of Drinking Soda Pop | Dherbs, Dherbs News, Alternative Medicine, Natural Remedies, Herbal Remedies, Herbal Supplements, Natural Remedies, Soda Pop, Is Soda Bad for You." Web. 8 Apr. 2015.

Donvan, John. "Meet First Child Ever Diagnosed with Autism." *ABC News*. N.p. 2010. Web.

Dr. E.H. Bronner Mfg. Research Chemist, Los Angeles, SOTT.net, "Small Amounts Fluoride Destroy the Will to Resist—Sott.Net." N.p.

Dr. Sircus. "Iodine and Detoxification." N.p. 2009. Web.

Dray, Tammy. "Health Effects of Yellow 5 Food Coloring." *Livestrong*.com. N.p. 2015. Web.

Edwards, Kasey. "When Birth Control Pills Become Dangerous." *Daily Life*. N.p. 2013. Web.

GMO Awareness. "GMO Defined." N.p. 2011. Web.

Geib, Aurora. "GMO Alert: Top 10 Genetically Modified Foods to Avoid Eating." *NaturalNews*. N.p. 2012. Web.

Green, Nastassia. "Moderation Is Health! High Oleic Sunflower Oil Adverse Effects." Oilypedia.com - *Benefits and Uses Of Supplemental and Essential Oils*. Web.

Group III, DC, ND, DACBN, DCBCN, DABFM, Dr. Edward. 'The Hidden Formaldehyde in Everyday Products." *Dr. Group's Natural Health & Organic Living Blog*. N.p. 2009. Web.

Gutierrez, David. "Teflon and Related Chemicals Linked to Arthritis." *NaturalNews*. N.p. 2013. Web.

Guyenet, Stephan. "Whole Health Source: Seed Oils and Body Fatness— A Problematic Revisit." *Wholehealthsource.blogspot*.com. N.p.

Hoffman, Dr. Ronald. "Estrogen Dominance Syndrome." *Drhoffman*.com. N.p. 2013. Web.

James, Susan. "What's in Your Beer? Fish Bladder and Antifreeze Ingredient?" *ABC News*. N.p. 2014. Web.

Josephson, Julian. "Chemical Exposures: Prostate Cancer and Early BPA Exposure." *Environmental Health Perspectives* 114.9 (2006): A520. Web.

Kindy, Kimberly. "Are Secret, Dangerous Ingredients in Your Food?" *Washington Post*. N.p. 2014. Web.

Larsson, Eva. "What is Oxybenzone in Sunscreen? Danger, Concerns, & Safety." *Morenature*.com. N.p. 2013. Web.

Livni, Ephrat. "Danger of Toxic Chemical in Cosmetics?" *ABC News*. Web.

Lord, Joel. "VRM: Health Matters Part 1 « Vaccine Resistance Movement." *Vaccineresistancemovement*.org. N.p. 2010. Web.

Lurie, Julia. "Your Bottled Water Comes from the Most Drought-Ridden Places in the Country." *Mother Jones*. N.p. 2014. Web.

Luther, Daisy. "Mysterious Deaths in Alabama: Could They Be Related to Monsanto's Bt Cotton Crops?" *Theorganicprepper*.ca. N.p. 2013. Web.

Malone, Luke. "The Cancer Danger in Canned Food." *The Sydney Morning Herald*. N.p. 2011. Web.

McCarthy, April. "Prevent Disease.Com - Organic Apples Beat Conventionals on Antioxidants." *Preventdisease*.com. N.p. 2009. Web.

McCarthy, April. "Study Shows How Organic Tomatoes Exceed Conventional in Antioxidant Value." *Preventdisease*.com. N.p. 2012. Web.

Mercola.com. "Bottled Water and Its Health and Environmental Impact." N.p. 2011. Web.

Monica and Dennis, RoseOfSharonAcres.com

Monsantoblog.com. "What's Served in Monsanto's Cafeterias? | Beyond the Rows." N.p. 2012. Web.

Myth Busting. ButterBeliever. "Think Fat-Free Milk is Healthy? 6 Secrets You Don't Know About Skim | Butter Believer." *Butterbeliever*.com. N.p. 2012. Web.

Nationalfork.com. "Search Results Prisons Are Poisoning Inmates with Soy | The National Fork." N.p. 2011. Web.

NaturalNews. "NBC News Declares 'Billions Could Starve' As America's Water Aquifers Run Dry." N.p. 2014. Web.

Newhealthguide.org. "High Estrogen in Men | New Health Guide." Web.

O'Brien, Robyn. "Organic Food Vs. Conventional: What the Stanford Study Missed." *The Huffington Post.* N.p. 2012. Web.

October 2009.KidsWithFoodAllergies.org, How to Read a Label for Soy Allergy

Ogburn, Stephanie. "The N2 Dilemma: Is America Fertilizing Disaster?." *Grist.* N.p. 2010. Web.

Organicapoteke.com. "Natural Skin Care : The Dangers of Phenoxyethanol | Organic Apoteke." Web.

Osansky, Dr. Eric. "3 Common Causes of Estrogen Dominance." *Naturalendocrinesolutions*.com. Web.

O'Shea, Dr Tim. "Cancer and Chemotherapy Epidemic | The Doctor Within.." *Thedoctorwithin*.com. Web.

O'Shea, Dr Tim. "Vaccines and the Peanut Allergy Epidemic | *The Doctor Within*." Thedoctorwithin.com. Web.

Park, Alice. "Breaking News, Analysis, Politics, Blogs, News Photos, Video, Tech Reviews - TIME.Com." *TIME*.com. N.p. 2008. Web.

Paull, Dr. John. "Journal of Organic Systems." *Journal of Organic Systems* (2013): n. pag. Web.

Philpott, Tom. "USDA Moves To Let Monsanto Perform Its Own Environmental Impact Studies On Gmos." *Grist.* N.p. 2011. Web.

Rattue, Petra. "Autism Linked to Industrial Food or Environment." *The Huffington Post*. N.p. 2012. Web.

Reinagel, MS. LD/N, CNS, Nutrition Diva, Monica. "Is Powdered Milk Bad for You?" *Quick and Dirty Tips*. N.p. 2010. Web.

Rettner, Rachael. "What is Estrogen?" *LiveScience*.com. N.p. 2014. Web.

Sott.net. "Scientists Cure Cancer, But No One Takes Notice—Sott.Net." N.p. 2012. Web

Sarah TheHealthyHomeEconomist. "Natural' Sodas Made with GMO Sugar Scam Consumers | The Healthy Home Economist." *The Healthy Home Economist*. N.p. 2013. Web.

Sears, Dr. William. "Ask Dr. Sears: Cows' Milk for Babies?" *Parenting*. Web.

Sifferlin, Alexandra. "Skim Milk is Healthier Than Whole Milk, Right? Maybe Not | TIME.Com." *TIME*.com. N.p. 2013. Web.

Stewardship, Katie. "The Real Story of Homogenized Milk, Powdered Milk, Skim Milk, and Oxidized Cholesterol." *Kitchen Stewardship | A Baby Steps Approach to Balanced Nutrition*. N.p. 2010. Web.

Sullivan, Kevin M., and Dr. Edward F. Group III, DC, ND, DACBN, DCBCN, DABFM. "The Interaction of Agricultural Pesticides and Marginal Iodine Nutrition Status as a Cause of Autism Spectrum Disorders." *Environ Health Perspect* (2015): n. pag. Web.

Talty, B.S.Ed. M.A, Caryn. "Do GMO Foods Cause Autism? Read About the GMO Crops Autism Connection." *Healthy Family*. N.p. 2012. Web.

Tarr Kent, Linda. "Dangers of Sucralose | Livestrong.com." Livestrong.com. N.p. 2013. Web.

Terpstra, Aliss. "Types of Fluoride." *Fluoride Detective*. N.p. 2012. Web.

Frye, Patrick. The Inquisitr News. "California Drought Leads to Catastrophic California Earthquake Predictions in 2014." N.p. 2014. Web.

Therapies for Autism, Gastrointestinal, and Neurological Disorders. 1st ed. research.va.gov, 2015. Web.

Tsukamura, Midoriko et al. "Dietary Maltitol Decreases the Incidence of 1,2-Dimethylhydrazine-Induced Cecum and Proximal Colon Tumors in Rats." *The Journal of Nutrition* 128.3 (1998): 536-540. Web.

Vani Hari "How to Eat Organic on a Budget." Food Babe. N.p. 2013. Web.

Webmd.com. "A History of Autism." Web.

Weitzlux.com. "Learn More about the Dangers Behind Dibutyl Phthalate Exposure." Web.

Wind, Rebecca. "Many American Women Use Birth Control Pills for Noncontraceptive Reasons." *Guttmacher*.org. N.p. 2011. Web.

Wood, Fred. "Health Effects of Yellow 5 Food Coloring | *Ehow*." eHow. Web.

Wright, Carolanne. "Political and Corporate Elite Shun GM Food on Their Own Plate." *NaturalNews*. N.p. 2012. Web.

Zerbe, Leah. "Cancerous Shampoo?!" *Prevention*. N.p. 2013. Web.

Zielinski, Eric L. "Sorbitol Causes Premature Cataracts, Retinopathy, Heavy Weight Loss, and Peripheral Neuropathy." *NaturalNews*. N.p. 2012. Web.

Zonis, Stephanie. "Organic Matter Archive - The Nibble Gourmet Food Magazine." *Thenibble*.com. N.p. 2005. Web.

www.ingramcontent.com/pod-product-compliance
Lightning Source LLC
Chambersburg PA
CBHW070226190526
45169CB00001B/101